D1641662

Chemical and Enzymatic Synthesis of Gene Fragments

A Laboratory Manual

Edited by H. G. Gassen and Anne Lang

Chemical and Enzymatic Synthesis of Gene Fragments

A Laboratory Manual

Edited by H. G. Gassen and Anne Lang

Weinheim · Deerfield Beach, Florida · Basel · 1982

Prof. Dr. Hans G. Gassen
Dipl.-Chem. Anne Lang
Institut für Organische Chemie und Biochemie
der Technischen Hochschule
Petersenstr. 22
D-6100 Darmstadt

Publisher's editor: Hans F. Ebel

Production manager: Peter J. Biel

This book contains 69 figures and 15 tables

Deutsche Bibliothek Cataloguing-in-Publication Data

Chemical and enzymatic synthesis of gene fragments:
a laboratory manual / ed. by Hans G. Gassen and Anne Lang. –
Weinheim; Deerfield Beach, Florida; Basel:
Verlag Chemie, 1982
ISBN 3-527-26063-3
NE: Gassen, Hans G. [Hrsg.]

Printing: betz-druck gmbh, D-6100 Darmstadt 12
Bookbinding: Josef Spinner GmbH, D-7583 Ottersweier
Printed in the Federal Republic of Germany

Preface

In the past the chemical and enzymatic synthesis of oligonucleotides of defined sequence had to be left to a few experts. Now, however, with the triester approach, the phosphite method and the solid-support techniques gene fragment synthesis has turned into an easy procedure even for a non-chemist. Due to the elegant chemistry involved, all methods work without sophisticated equipment and are prone to mechanisation and eventual automation. It is hoped that combined chemical-enzymatic gene synthesis may become a standard technique in a molecular biology laboratory, such as DNA sequencing or in-vitro recombination of nucleic acids.

We omitted chemical RNA synthesis, since this field is developing so rapidly at the moment that one has to refer to the original publications. However, we included enzymatic synthesis of RNA fragments, procedures which already have obtained a high degree of standardisation.

Most of the contributions are revised versions of the protocols supplied for the EMBO sponsored course on "Automated Chemical and Enzymic Gene Synthesis", held in Darmstadt, March 21 to April 3, 1982. The protocols were improved on the basis of the experience of 30 student scientists with chemical, biological or medical backgrounds. Previously omitted procedures, such as the wandering spot method for oligonucleotide analysis, were included. In editing the manuscript we encountered problems with the nomenclature of nucleic acid components. In unambiquous cases we favoured a simple description, hoping for example, that oligodeoxynucleotide is always understood to mean oligo-2′-deoxyribonucleotide.

This book aims to provide those interested in DNA/RNA research with state-of-the-art methods in the synthesis, purification, and analysis of DNA and RNA fragments. The editors wish to thank the authors for their efforts in preparing manuscripts from the revised laboratory protocols. We gratefully acknowledge the skill and the patience of Mrs. E. Rönnfeldt in typing the manuscripts.

We express our thanks to Verlag Chemie for the friendly and very efficient cooperation.

Darmstadt, in July 1982

H.G. Gassen
A. Lang

Contents

Chemical Synthesis of Polydeoxyribonucleotides of Defined Sequence

Biochemical Modification of Gene Fragments

List of Contributors

Blöcker, Helmut see Frank, R.

van Boeckel, C.A.A. see van Boom, J.H.

van Boom, Jacques H., G.A. van der Marel, C.A.A. van Boeckel, G. Wille, and *C.F. Hoyng* (see extra entry)
Department of Organic Chemistry
Gorlaeus Laboratories
University of Leiden
P.O. Box 9502
2300 RA Leiden
The Netherlands

Caruthers, Marvin H.
Department of Chemistry
University of Colorado
Boulder, CO 80309
U.S.A.

Cook, Ronald M., D. Hudson, E. Mayran, and *J. Ott*
Biosearch
1281-F Andersen Drive
San Rafael, CA 94901
U.S.A.

Dörper, T. see Winnacker, E.-L.

Eckert, V., see also Ohtsuka, E.
Fachgebiet Biochemie
Technische Hochschule Darmstadt
Petersenstr. 22, D-6100 Darmstadt
Federal Republic of Germany

Eiband, J. see Seliger, H.

Eick, D. see Fritz, H.J.

Frank, Ronald, and *H. Blöcker*
GBF, Gesellschaft für Biotechnologische Forschung mbH
Mascheroder Weg 1, D-3300 Braunschweig
Federal Republic of Germany

Fritz, Hans J., W.-B. Frommer, W. Kramer, and *W. Werr*
Institut für Genetik, Universität zu Köln
Weyertal 121, D-5000 Köln 41
Federal Republic of Germany

Fritz, Hans J., D. Eick, and *W. Werr*
Institut für Genetik, Universität zu Köln
Weyertal 121, D-5000 Köln 41
Federal Republic of Germany

Frommer, Wolf-Bernd see Fritz, H.J.

Gait, Michael J., H.W.D. Matthes, M. Singh, B.S. Sproat, and *R.C. Titmas*
Laboratory of Molecular Biology
Medical Research Council Centre
University Medical School
Hills Road, Cambridge, CB2 2QH
U.K.

Gassen, Hans G. (several contributions), *V. Eckert, C. Gatz, W. Hillen, G. Klock, A. Lang,* and *M. Schmitt*
Fachgebiet Biochemie
Technische Hochschule Darmstadt
Petersenstr. 22, D-6100 Darmstadt
Federal Republic of Germany

Gatz, Christiane, and *W. Hillen*
Fachgebiet Biochemie
Technische Hochschule Darmstadt
Petersenstr. 22, D-6100 Darmstadt
Federal Republic of Germany

Hauel, N. see Seliger, H.

Hillen, Wolfgang
Fachgebiet Biochemie
Technische Hochschule Darmstadt
Petersenstr. 22, D-6100 Darmstadt
Federal Republic of Germany

Hoyng, C.F.
Genentech, Inc.
South San Francisco, CA 94080
U.S.A.
(*see also* van Boom, J.H.)

Hudson, D. see Cook, R.M.

Klein, S. see Seliger, H.

Klock, Gerd
Fachgebiet Biochemie
Technische Hochschule Darmstadt
Petersenstr. 22, D-6100 Darmstadt
Federal Republic of Germany

Kramer, W. *see* Fritz, H.J.

Krusche, Jörg U., see also McLaughlin, L.W.
Dupont de Nemours (Deutschland) GmbH
Abteilung Analytische Instrumente
Dieselstr. 18, D-6350 Bad Nauheim
Federal Republic of Germany

Lang, Anne, and *H.G. Gassen*
Fachgebiet Biochemie
Technische Hochschule Darmstadt
Petersenstr. 22, D-6100 Darmstadt
Federal Republic of Germany

van der Marel, G.A. *see* van Boom, J.H.

Matthes, H.W.D. *see* Gait, M.J.

Mayran, E. *see* Cook, R.M.

McLaughlin, Larry W., and *J.U. Krusche*
(see extra entry)
Max-Planck-Institut für experimentelle Medizin
Hermann-Rein-Str. 3, D-3400 Göttingen
Federal Republic of Germany

Narang, C.K. *see* Seliger, H.

Ohtsuka, Eiko, and *V. Eckert*
(see extra entry)
Faculty of Pharmaceutical Sciences
Osaka University
1–6 Yamadaoka, Suita
Osaka, Japan 565

Ott, J. *see* Cook, R.M.

Schmitt, Marion, and *H.G. Gassen*
Fachgebiet Biochemie
Technische Hochschule Darmstadt
Petersenstr. 22, D-6100 Darmstadt
Federal Republic of Germany

Schott, Herbert
Institut für Organische Chemie
Universität Tübingen
Auf der Morgenstelle 18, D-7400 Tübingen
Federal Republic of Germany

Seemann-Preising, B. *see* Seliger, H.

Seliger, Hartmut, S. Klein, C.K. Narang, B. Seemann-Preising, J. Eiband, and *N. Hauel*
Sektion Polymere, Universität Ulm
Oberer Eselsberg, D-7900 Ulm
Federal Republic of Germany

Singh, M. *see* Gait, M.J.

Sproat, B.S. *see* Gait, M.J.

Titmas, R.C. *see* Gait, M.J.

Uhlenbeck, Olke C.
Department of Biochemistry
School of Chemical Sciences
University of Illinois
Urbana, Illinois 61801
U.S.A.

Werr, W. *see* Fritz, H.J.

Wille, G. *see* van Boom, J.H.

Winnacker, Ernst-Ludwig, and *T. Dörper*
Institut für Biochemie der Ludwig-Maximilians-Universität
Karlstr. 23, D-8000 München 2
Federal Republic of Germany

SYNTHESIS OF OLIGODEOXYRIBONUCLEOTIDES BY A CONTINUOUS FLOW, PHOSPHOTRIESTER METHOD ON A KIESELGUHR/POLYAMIDE SUPPORT

Michael J. Gait, Hans W.D. Matthes, Mohinder Singh, Brian S. Sproat, and Richard C. Titmas

Laboratory of Molecular Biology
Medical Research Council Centre
Cambridge

SUMMARY

A new composite polydimethylacrylamide/kieselguhr support is used in a continuous flow assembly apparatus for the efficient solid phase synthesis of a deca- and a nonadecadeoxyribonucleotide using phosphotriester intermediates.

1 Introduction

There are few scientists these days interested in DNA who have not yet heard of solid phase synthesis of DNA (if nowhere else in advertisements for DNA synthesis machines in Nature). In 1975 when Hartmut Seliger gave an EMBO Course entitled "Synthesis of oligonucleotides on polymer-supports" it was a different story. Then few DNA chemists gave solid phase synthesis much chance of success and few would contemplate its use beyond tri- or tetranucleotides. Automated synthesis was merely a distant dream.

However, the principle of solid phase synthesis has been recognised since its inception in the mid-1960s [1] as an extremely elegant one. It is particularly suitable for preparation of biopolymers of defined sequence such as oligonucleotides and peptides. In essence one end of a growing chain is protected by an insoluble, macromolecular protecting group, whilst the other end is reacted with the next unit to be coupled (i.e. Ⓟ -A-B + C-x ⟶ Ⓟ -A-B-C-x). Any excess of C-x can be removed by mere filtration and washing. After selective removal of protecting group x the process can be repeated by addition of the next unit D-x and so on (Ⓟ -A-B-C-X ⟶ Ⓟ -A-B-C + D-x ⟶ Ⓟ -A-B-C-D-x). At the end of the synthesis the bond between the chain and support Ⓟ is broken and the chain is released into solution and purified by chromatography.

Whereas by the mid-1970s short peptides in many cases could be assembled well, and indeed commercial peptide synthesis machines were already available, the coupling of nucleotides on solid phase did not seem to be as efficient as in conventional solution reactions, such that yields of desired products became vanishingly small. A full discussion of the various polymer-supports and methods tried can be found in some recent review articles [2]. It is sufficient here to state that there were two basic reasons for these failures: 1. the supports chosen were not appropriate to the chemical reactions used for DNA synthesis and 2. the chemical reactions used were insufficiently selective and reliable to meet the two extra demands of the solid phase approach.

First these are heterogeneous reactions which are liable to be slower than their solution counterparts. The minimum increase in $t_{1/2}$ is probably about 2-3 times under optimum conditions but this can be much worse if there is any incompatability between support, solvent and chemistry. Thus the solid phase method works best for relatively fast reactions ($t_{1/2}$ measures in sec or a few min), where small rate changes do not significantly increase the assembly time and can be easily allowed for. Secondly, it is common to push reactions to completion on solid phase by use of a substantial excess of added nucleotide in solution (i.e. obtaining essentially pseudo-first order kinetics). In this situation the slightest reactive impurity in the nucleotide in solution gets magnified by virtue of the excesses used (e.g. 1% impurity in a nucleotide used in 10-fold excess over growing chain can give up to 10% side reaction).

The classical phosphodiester approach to DNA synthesis, despite some considerable success in conventional solution methods, turned out not to be particularly suited to solid phase, even though many attempted it. Partly this was due to poor choice of polymer supports. For example, polystyrene gel is inappropriate because its non-polar nature does not match the highly polar nature of oligonucleotide chains with their negatively charged phosphodiesters. In addition phosphodiester chemistry is not sufficiently selective to give consistently high yields in coupling reactions, an important necessity for successful solid phase synthesis, and monomer units were difficult to obtain in high purity. Even so, some reasonable syntheses of oligodeoxyribonucleotides up to about 12 units were obtained using in particular polyamide supports [3,4].

Phosphotriester chemistry (including the companions "phosphite-triester" and "phosphoroamidite") is much more suitable for solid phase synthesis and in the last two years or so has proved to be highly successful. The oligonucleotide chains as they are assembled contain uncharged, fully protected phosphates and are found to be compatible in solvation properties with a variety of polymer supports (polyamide, polystyrene, silica). Also coupling reactions are much more selective and give higher yields.

2 EXPERIMENTAL OUTLINE

This contribution is designed to be an introduction to all the important techniques necessary to make oligodeoxyribonucleotides by the phosphotriester approach on solid phase. The methods described are those in current use in the author's laboratory and maximises the use of commercially available materials and equipment.

Two syntheses will be described: the decanucleotide, d(C-C-G-A-T-A-T-C-G-G). This is a self-complementary sequence that has a central six nucleotides which are the recognition sequence for the restriction endonuclease EcoRV. The oligonucleotide is wanted for an nmr study of its interaction with the endonuclease and also for crystallisation and X-ray analysis. The synthesis will demonstrate the use of monomer units in couplings to solid phase on moderate scale (10 μmol) for the preparation of a few mg of decanucleotide.

The nonadecanucleotide, d(T-G-G-T-C-A-T-A-G-C-T-G-T-T-T-C-C-T-G): this sequence is complementary to a region of the single-stranded bacteriophage DNA M13 upstream of the normal DNA insertion sites used in the cloning and sequencing procedures of Sanger's group. The oligonucleotide is designed as a primer for the preparation of partial duplexes of M13 that leave inserted DNA as essentially single-stranded. This can be used as a hybridisation probe for the selection of new clones containing oppositely orientated inserts and to obtain overlap information that would otherwise be difficult to obtain. The synthesis will be on small scale (5 μmol) and will demonstrate the use of dinucleotide blocks as coupling units.

3 DISCUSSION OF METHODS

3.1 Polymer support and its functionalisation

3.1.1 The support

Most recently published methods of DNA synthesis by the phosphotriester route have used gel resins of either polystyrene [5], polyacryloylmorpholide [6] or our own polydimethylacrylamide copolymers [7,8,9]. These are solvent-swollen polymers that are generally handled in glass reaction vessels where solvents are added batchwise, followed by some sort of agitation and then filtration by application of partial vacuum or nitrogen pressure. Such techniques, although perfectly acceptable, suffer from the disadvantage that because of solvent hold-up in the gel, several washings are needed to remove a previous solvent or reagent. This batchwise technique is cumbersome to automate and can be time-consuming and expensive on solvents.

The most efficient way to wash a polymer-support is to pack it into a column and pass solvents through continuously. Unfortunately low crosslinked, gel resins, such as those suitable for synthesis, tend to pack down under pressure of solvent leading to irregular flow characteristics and high back pressure, and they are therefore unsuited to this approach. Porous inorganic materials, such as silica, can be used for solid phase synthesis by phosphotriester [10] or phosphite [11] methods. However, there is still tendency, in our experience, to generate back pressures especially with the finer mesh varieties, leading to inconsistent flow and other technical problems. It is also difficult to obtain reproducible nucleoside loadings on silica and care must be taken to cap off unfunctionalised silanol groups.

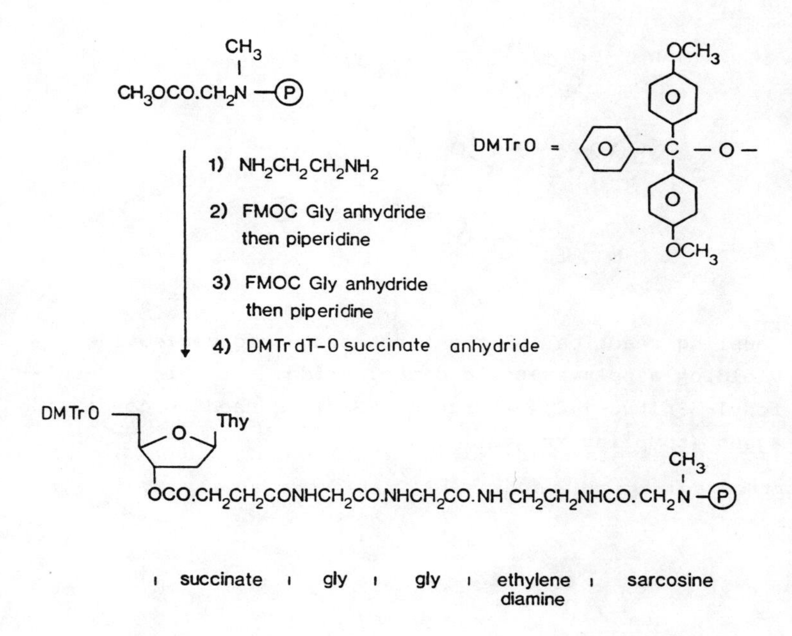

Figure 1 Functionalisation of polydimethylacrylamide/kieselguhr

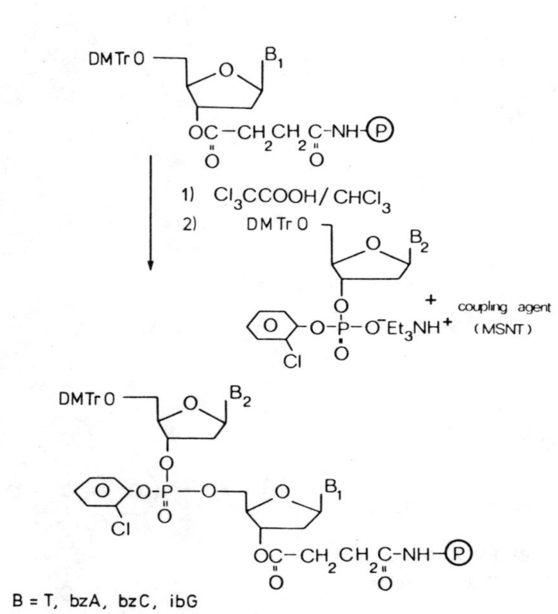

B = T, bzA, bzC, ibG

Figure 2 Coupling reaction for the first protected nucleotide yielding a polymer-bound dinucleotide. Mesitylene-2-sulfonyl-3-nitro-1,2,4-triazole (MSNT) is used as condensing agent (coupling reagent)

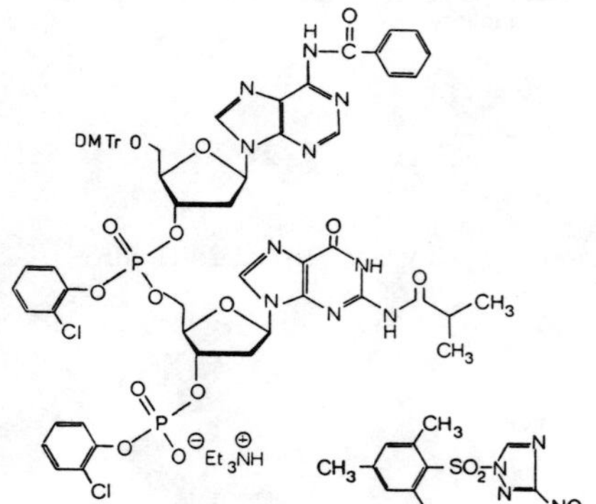

The coupling agent MSNT

Figure 3
Structure of the fully protected dimer d(ApG) and the coupling reagent mesithylene-2-sulfonyl-3-nitro-1,2,4-triazole

Recently a new composite support has been developed consisting of fabricated beads of inorganic material (e.g. kieselguhr) that contain large pores, several 100 nm in diameter, in which polydimethylacrylamide gel resin has been prepared in situ [12]. The support combines the excellent flow characteristics of the beads, that generate negligible back pressures even at fairly high flow rates, with the proven value of the polyamide gel. Rapid diffusion of solvent takes place in and out of the beads so that no agitation of the material is required during synthesis. The material has been used successfully in flow systems for both peptide [12] and oligonucleotide synthesis [13] and is now commercially available (LKB Biochrom, Ltd., Cambridge, U.K. and Victor Wolf, Ltd, Clayton, Manchester, U.K.).

3.1.2 Functionalisation

The support contains a controlled amount of ester functionalities introduced in the polymerisation reactions used in its preparation as acryloyl sarcosine methyl ester. Batches of support are assayed by amino acid analysis for sarcosine content. This represents the nominal loading value of nucleoside obtainable for the particular batch. So far batches containing ca 0.1 $mmol \cdot g^{-1}$ sarcosine have been used and through a series of precycling reactions (Fig. 1) nucleoside loadings of 85 - 90 $\mu mol \cdot g^{-1}$ have been consistently obtained. The support is treated with ethylene diamine overnight, washed well with dimethylformamide (DMF) and then coupled under standard peptide coupling conditions with a protected glycine derivative (Fmocgly) prepared as its symmetrical anhydride. The fluorenylmethoxycarbonyl (Fmoc) protecting group is removed by treatment of the support with piperidine/DMF and a second protected glycine derivative added as before. The double glycine spacer has been found necessary in order to obtain consistent rates of internucleotidic coupling in DNA assembly.

To attach the first nucleoside the support is once again treated with piperidine/DMF and then coupled with the symmetrical anhydride derivative of one of the four 5'-O-dimethoxytrityl-2'-deoxynucleoside-3'-O-succinates in an analogous way to a peptide coupling reaction.

3.2 Assembly of oligonucleotide chains

Oligonucleotides are synthesised from 3' to 5' ends in a series of assembly cycles involving only two chemical reactions per cycle (Fig. 2). In common with other solid phase methods the extremely acid labile dimethoxytriphenylmethyl (dimethoxytrityl) group is used for protection of the 5' position (the pixyl [14] protecting group is an acceptable alternative). Of all the protic acids which can be used for deprotection we have found that trichloroacetic acid (TCA) is the best [8]. Oligonucleotide chains containing deoxyadenosine are susceptible to slight depurination (loss of adenine) under acidic conditions, and upon removal of protecting groups at the end of the synthesis the chain fragments at the depurinated site. We have observed, however, that depurination is negligible in most cases (much less than 1% per deprotection step per adenine residue) using brief TCA treatment for the removal of terminal dimethoxytrityl groups. The alternative reagent, zinc bromide, which gives rise to little or no depurination, has been used in conjunction with polystyrene [15] and silica supports [16]. Unfortunately this reaction is prone to inhibition by other chelating agents including amino groups of amide bonds. Increasingly slow reactions have been observed particularly for highly functionalised resins [17] as the oligonucleotide chain is extended and this requires careful monitoring. As might be anticipated zinc bromide deprotects too slowly when used in conjunction with polyamide supports [17,18]. TCA is not subject to such inhibition and gives consistent deprotection within 3 min.

The acidic deprotection step generates free hydroxyl groups on the support, which are coupled in a second step with an appropriately protected monomer or dimer unit (Fig. 3). These contain dimethoxytrityl groups on the 5' position, acyl groups protecting the exocyclic amino groups of adenine, cytosine and guanine, and 3'-O-2-chlorophenyl phosphodiesters. These latter moieties react with hydroxyl groups to give fully protected phosphotriesters. The dehydration (coupling) reaction is carried out in pyridine with the coupling agent mesitylene-2-sulphonyl-3-nitro-1,2,4-triazole (MSNT) [19]. The two reactions of deprotection and coupling are separated by appropriate polymer washing steps. In contrast to other published solid phase methods we find no particular benefit in incorporating a third reaction of treatment of the support with a capping reagent after coupling.

The polymer support is contained in a small glass column connected to a manually operated solvent delivery system operated by slight pressure of an inert gas such as argon or nitrogen. All fittings need only be low pressure of Teflon and glass. An inexpensive kit is commercially available (Omnifit, Ltd., Cambridge, U.K.), but any reliable flow system, high performance liquid chromatography (hplc) or automated synthesiser could potentially be adapted for the purpose. The cycle time is 60-80 min and involves very little manual work.

3.3 Deprotection

After the appropriate number of assembly cycles three stages of deprotection are required but without intermediate purifications.

(1) The support is treated with 1,1,3,3-tetramethylguanidinium-syn-2-nitrobenzaldoximate [20] in aqueous dioxan to remove 2-chlorophenyl protecting groups from internucleotide phosphates and to break the succinyl linkage and liberate oligonucleotides into solution. The use of this oximate reagent gives rise to very little unwanted cleavage of internucleotide bonds, which for many years was a major side reaction resultant from the use of inferior reagents for this step.
(2) The oligonucleotide is treated with concentrated ammonia solution in a sealed tube at 50°C for 5 h to remove acyl protecting groups.
(3) The final step is treatment with acetic acid:water (8:2, v/v) for 30 min to remove terminal dimethoxytrityl groups.

3.4 Purification by high performance liquid chromatography (hplc)

hplc is recommended for most oligonucleotides up to 20 units and when the highest purity is required. Initial purification is best carried out by ion exchange chromatography which separates oligonucleotides principally (but not exclusively) on the basis of their length. Oligonucleotides are polyanions having formal negative charges increasing integrally as a function of chain length. Most likely impurities in solid phase synthesis are either truncated (prematurely terminated) or failure sequences (missing one or more units), or arise from chain cleavage reactions. Such impurities will be shorter in length and hence are less strongly retained on ion exchange columns. This makes identification of the desired peak on chromatograms usually straightforward. For longer chains or those which have some self-complementarity or are rich in guanine, disaggregants (ethanol or formamide) and elevated temperatures are necessary in this chromatography step.

After gel filtration to remove salts a second purification step can be carried out by reversed phase chromatography. Here separation relies on differences in hydrophobicity and particularly good resolution can be obtained of minor impurities with the same or similar length (e.g. base modified chains). Oligonucleotides having gone through both chromatography steps are reproducibly 99% pure or better as determined by analysis of ^{32}P-labelled samples. The second step (reversed phase) can often be omitted for the shorter chains.

3.5 Purification by polyacrylamide gel electrophoresis (page)

This technique can be used in place of ion exchange hplc as a first line purification method. It is particularly useful for long chains which cannot be resolved by hplc or for purification of identical length chains synthesized as a mixture for gene probes. Once again separation occurs primarily as a function of chain length but products of mixed syntheses give less sharp bands on the gel as those of single syntheses as visualised under the uv-lamp. Oligonucleotides can be recovered in good yield by soaking gel slices in water. Buffer and other gel contaminants are removed by a simple desalting procedure.

4 Experimental section

Experimental outline:

The major techniques covered in this course are divided into three stages: (1) Functionalisation of support; (2) assembly; and (3) deprotection and purification. The experiments are designed for a period of about 5 days for a minimum of 2 people and assumes that solvents and reagents (including nucleotide derivatives) have been prepared in advance.

A full list of materials (both chemicals and equipment, together with suggested suppliers) can be found in appendix 1.

A suggested timescale is as follows:

Day 1	Experiment 4.1:	Complete.
	Experiment 4.2.1:	Set up bench machine.
	Experiment 4.2.2:	Check flow rates and establish cycle time.
	Experiment 4.2.3:	Weigh out aliquots of monomers, dimers and coupling agent, begin co-evaporations.
	Experiment 4.2.4:	Weigh out support from Experiment 4.1 and pre-equilibrate in column.
Day 2	Experiment 4.2.3: (continue)	Complete co-evaporations of monomers and dimers.
	Experiment 4.2.5:	Begin assembly (4-6 cycles).
Day 3	Experiment 4.2.5: (continue)	Complete assembly.
	Experiment 4.2.6:	Resin drying and oximate treatment (overnight).
Day 4	Experiment 4.2.6: (continue)	Complete deprotection.

	Experiment 4.2.7.1:	Run hplc standards. Ion exchange hplc purification.
	Experiment 4.2.7.3:	Desalt samples for gel purification.
Day 5	Experiment 4.2.7.2: (continue)	Biogel P2 desalting. Reversed phase hplc.
	Experiment 4.2.7.3: (continue)	Analytical gel.

4.1 Functionalisation of support

Kieselguhr/polyamide support as it comes at present from the supplier requires overnight treatment with ethylene diamine followed by three cycles of peptide-type coupling reactions, two with Fmoc glycine and one with a deoxynucleoside-3'-O-succinate derivative. The cycles involve very similar manipulations and therefore in order to save time this experiment will start with support that has already been carried through the first two cycles. The full procedure for support functionalisation is given in Appendix 2.

Weigh out 1 g of support provided into the glass reaction vessel (see Appendix 2). Add DMF (5 mL), stopper and swirl gently for 1 min (N.B. do not shake the support vigorously or continuously for more than a 0.5-1 min period as this can be abrasive causing formation of fines which would eventually block the glass sinter). Allow to settle and leave the support to swell for 45-60 min.

Treat the support with piperidine:DMF (2:8, v/v) for 10 min and then 5 washes with DMF (2 min each). Using gentle pressure of air or an inert gas blow out the liquid through the sinter into a waste container. After each solvent addition (5 mL) swirl gently for 0.5-1 min and allow to stand for the remaining period, followed by filtration.

At this point take out a few beads of wet support with a spatula and put into a small sintered glass funnel for later testing.

Weigh out 0.41 g (0.5 mmol) of 5'-O-dimethoxytrityl-2-N-isobutyryl-2'-deoxyguanosine-3'-O succinate (preparation procedure in Appendix 2) and dissolve in 3 mL of methylene chloride in a small stoppered flask. Make up a solution of dicyclohexylcarbodiimide (DCC, 51 mg, 0.25 mmol) (CARE - use eye protection and gloves) in 1 mL of methylene chloride and add this to the nucleoside succinate solution. Put in a small Teflon stirring bar and stopper the flask. Stir on a magnetic stirrer for 20 min when a precipitate should be seen. Filter the mixture through a small sintered glass funnel into a round bottomed flask. Evaporate to dryness on a rotary evaporator (no heating), take up in DMF (5 mL) and add the solution immediately to the support. Swirl the reaction vessel for 1 min and leave to stand for 1 h.

Filter support (N.B. nucleoside succinate can be recovered from the filtrate by evaporation and precipitation) and wash with five batches of DMF (2 min washes).

Take a few beads of wet support onto a small sintered glass funnel and wash (in parallel with previous sample above) with 5 times methylene chloride, 5 times dioxane and 5 times diethyl ether using Pasteur pipettes for each addition and a water pump if necessary to suck dry. The support samples should now be dry and granular.

Treat each of the samples as follows. Put a few granules into a small glass tube. Add one drop each of Ninhydrin Test Solutions 1, 2 and 3. Heat in an oven at 100°C for 5 min. A deep blue colour in solution and on the beads signifies the presence of free amino groups on the support. The second test sample should not contain free amino groups and should remain yellow or very faint purple. The test is extremely sensitive so that slight blue colours can be ignored. A deep blue colour in the second test sample suggests a failed reaction (probably due to incorrect preparation of nucleoside succinate anhydride as a result of poor quality DCC) and the cycle should be repeated.

If the test gives a good result, wash the body of the support with 5 times methylene chloride, 5 times dioxane and 5 times diethyl ether and dry in a desiccator for 30-60 min.

It is normal to assay support samples by amino acid analysis (sarcosine and glycine content) and trityl analysis.

4.1.1 Trityl analysis

Weigh accurately 2-3 mg of dry resin and put it into a 25 mL standard flask. Make up to the mark with 60% perchloric acid:ethanol (3:2, v/v) and shake for 5-10 min. Measure the absorbance of the solution at 495 nm in a spectrophotometer. Multiply this reading by 25 to give the total absorbance. Assuming 1 µmol of trityl group correspond to 71.7 absorbance units at 495 nm, calculate the total number of µmol of trityl group on the sample and hence the number of $\mu mol \cdot g^{-1}$. This value is a messure of the nucleoside loading of the support.

4.2 Assembly of oligonucleotide chains

This experiment has been set up for parallel synthesis of 2 oligonucleotides on 2 columns but using only 1 solvent delivery system. For this purpose the experiment is best operated by two individuals working alternately, i.e. one person carrying out a wash cycle on one column whilst the other column is at the stage of the coupling reaction. (An individual working through this experiment without instruction would be best advised not to attempt parallel synthesis the first time, although this is quite possible with a little practice.) The coupling reactions require 45-60 min whilst the washing cycle takes only about 20 min. So there should be no difficulty in accomplishing the two syntheses together. Although the scales of the two syntheses are different the washing cycle times are the same.

4.2.1 Assembly of bench synthesiser

A diagram of the system is shown in figure 4 and for the most part the assembly is fairly obvious. The following points should be noted: (1) Use the thick Teflon tubing (1.5 mm i.d.) for all argon filled lines and for delivery of the 4 solvents to the rotary valve. Use medium tubing (0.8 mm) for the column inlet and connect at least 1 m of thin tubing (0.3 mm) to the outlet side of the column. This causes a very slight backpressure enabling a reading to be obtained on the pressure gauge. It is normally necessary to cut the Teflon tubing to the required length and put on connectors at each end. Note that liquid flow connectors are all Teflon fittings whereas gas flow connectors can be made with Viton 'O' rings. Nitrogen can be used instead of argon if necessary. (2) Use the special long column (150 mm) to make a silica gel drying tube which is inserted into the argon delivery line as a precaution to prevent moisture getting into the system. Use the two short columns (50 mm) for the syntheses. Each synthesis column should have a septum fitting on the top, a Teflon sinter and variable length plunger at the bottom. (3) Put the 3-way valve on the inlet side of the columns after the rotary solvent selector valve. The two column outlets can be connected to just one waste bottle. (4) The rotary selector valve should have solvents connected to it in the following order (either clockwise or anti-clockwise): pyridine, chloroform, TCA, DMF, pyridine and then the sixth line blocked off (i.e. a stop position). The two pyridine inlets can come from one solvent bottle since there are three connectors on each bottle, two of which can be used as solvent outlets. (5) Before filling solvent bottles check for leaks in plumbing by putting some dichloromethane into each bottle, pressuri-

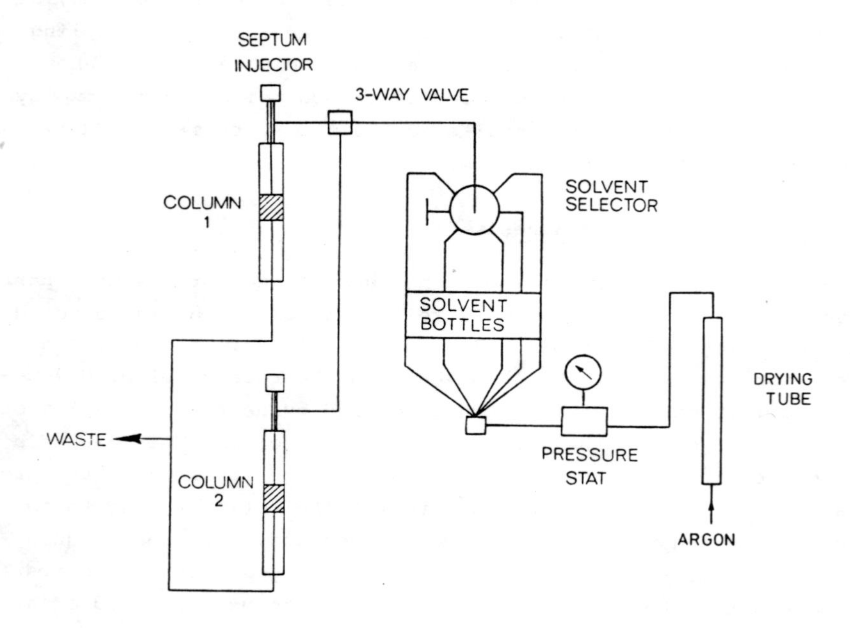

Figure 4 Sketchy drawing of a bench type synthesiser. For parts list see Appendix 1, No 14.

sing in turn and checking the flow through the entire system. Look particularly at liquid-liquid joints and tighten if necessary. Dry out bottles and solvent lines.

Flush all the bottles through with argon and then fill with solvents. DMF, pyridine and chloroform (but not TCA) should also have some fresh molecular 4 Å sieve beads added. The solvents should be predried as described in the appendix. (CARE. DMF is harmful to skin; pyridine and chloroform are also toxic.) TCA (10 g) should be from a fresh bottle, weighed out quickly and transferred immediately into the small (100 mL) bottle, chloroform (100 mL) added and the bottle quickly closed.
(6) To prevent particles of molecular sieve clogging the solvent lines hplc type stainless steel solvent prefilters should be inserted in the ends of pyridine, DMF and chloroform lines. (Note these are not provided at present with the Omnifit kits.)

4.2.2 Calibration of flow rates

Weigh out the required amount of polymer-support (unfunctionalised - for test only) into each column. Make sure the variable length plunger is at the outlet (bottom) of the column and the septum injector at the inlet (top). Fill the column with DMF from the septum inlet and allow the material to swell for a few min. Stir briefly to remove bubbles. Put on the septum and inlet line making sure that no bubbles of air are in the solvent line or column. Pass DMF through the system by opening the appropriate valves and by turning up the argon pressure to about 0.5-0.7 bar (8-10 psi). Adjust the pressure until the flow rate is 1.0 $mL \cdot min^{-1}$. In turn flow through the other solvents checking their flow rate too. Pyridine should have about the same flow rate, chloroform about double the flow rate. Do this for both columns.

4.2.3 Preparation of monomers, dimers and coupling agents

Full protocols for preparation of monomers and dimers in a form suitable for storage (at -20°C) is given in Appendix 3. These can also be purchased from a number of suppliers at varying cost (see Appendix 1) but little experience is yet available concerning the quality of these commercial materials. Monomers can be used for assembly up to perhaps 16-18 units. However, we have recently found that the use of barium salts (particularly of adenylic acid and guanylic acid) sometimes gives much poorer results than triethylammonium salts in assemblies (whereas both give satisfactory results in synthesis of dimers in solution). This is probably due to slightly higher levels of impurities in the barium salt procedures, which, as has been mentioned, is much more critical on solid phase. Even with triethylammonium salts the utmost care must be taken to ensure that only good quality materials are used in solid phase synthesis.

Dimers are recommended for assemblies of greater than 18 residues or for shorter chains when a smaller number of cycles is required. These are best stored as fully protected phosphotriesters. Terminal 2-cyanoethyl protection groups should be removed from the exact amount of dimer required by the procedure given in Appendix 3 preferably within a few days of being required. Precipitates are best collected by centrifugation and dried in vacuo in the centrifuge pots.

Calculate the total amount of each monomer required for the 10-mer synthesis.

Molecular weights, triethylammonium salts: thymidylate (T) = 836; adenylate (A) = 949; cytidylate (C) = 925; guanylate (G) = 933
i.e. T = 2 x 80 µmol, A = 2 x 80 µmol, C = 3 x 80 µmol, G = 2 x 80 µmol

These are the amounts you will need to weigh out for the synthesis into 4 flasks. One of the dimers (TG) occurs twice. 80 µmol (2 x 40 µmol) of this is provided whereas 40 µmol of the other dimers are provided.

In turn dissolve each monomer or dimer in a few mL of pyridine and transfer with a Pasteur pipette into a labelled pear-shaped 5 mL flask with B14 (small) ground glass joint (do not fill the flask more than one third full to prevent splashes). Connect to the rotary evaporator and evaporate to a solid (or gum). Use a water bath at 30°C. Add more pyridine and evaporate again. Do this about 4 times in all. Watch out for 'bumping' during the evaporation. Be sure to wash out the evaporator tube if this occurs. Quickly disconnect the flask from the evaporator and put on a silicone "Subaseal" septum. Make sure no water or moisture gets into the flask. (Some people like to let in dry argon to the evaporator before disconnecting the flask). Leave sealed flasks in a desiccator overnight.

4.2.4 Pre-equilibration of support

Clean out the columns from Experiment 4.2.2 and refill with weighed support obtained from Experiment 4.1 (60 mg for 19-mer synthesis, 120 mg for 10-mer synthesis). Add DMF to support, allow to stand for a few min and stir briefly to remove air bubbles. Connect septum to top of column making sure no air bubbles are in the line. Repeat for the second column. Flow DMF through both columns to ensure even flow.

4.2.5 Assembly of oligonucleotides

Assemble the 10-mer on column 1 and alternate on the machine assembling the 19-mer on column 2. Weigh into microvials MSNT (90 mg, 304 µmol per vial, for 10-mer synthesis; 60 mg, 202 µmol per vial for 19-mer synthesis) and seal with Teflon coated septa.

Check that no bubbles are present in either column and flow pyridine briefly through the system and select stop position in solvent selector.

Column drying, column 1 (10-mer synthesis): make up in a chemically inert 1 mL syringe a mixture of 0.1 mL phenyl isocyanate and 0.9 mL pyridine. Inject one column volume (0.4 mL) straight on to the column (N.B. slacken off the septum cap a quarter turn before piercing septum and then tighten before injection). Keep the syringe in the septum and inject 0.1 mL every 1 min until all the liquid is injected. Remove syringe slowly and leave column for 4 min.

Deprotection and wash, cycle 1: start the deprotection and wash cycle using the times indicated in the Table below. The times are suitable for both columns. After the final pyridine wash select stop position once again.

	Solvent	Time (min)
(1)	Pyridine	5
(2)	Chloroform*	3
(3)	10% TCA:chloroform (v/v)	3
(4)	DMF	5
(5)	Pyridine	5

*Due to the instability of pure chloroform that dichloromethane (distilled over phosphorous pentoxide) may be used instead.

Nucleotide addition, cycle 1: reactions are carried out using nucleotides dissolved in 0.6 mL of pyridine. Therefore dissolve the 'G monomer' in 2 times reaction volume of pyridine (there are two Gs to be added in the chain), i.e. 1.2 mL by injecting pyridine into the 'G flask'. When dissolved take up 0.6 mL into the syringe and inject into the MSNT vial (90 mg size). MSNT dissolves almost immediately. Once again take up the solution into the syringe. Inject this on to the column by the same procedure as for column drying with phenyl isocyanate except inject 0.35 mL first and then 50 µL every 1 min until all injected. (N.B. be sure to loosen the septum cap a little before piercing the septum.) Start counting time from the first injection point. Allow

60 min for the coupling so as to give the same cycle time as for column 2. Normally monomers only require 45 min and dimers 60 min. Because of the two column operation the total cycle time is best kept the same for the two columns. Wash out the syringe with pyridine. The remaining G monomer is used later in the synthesis. Switch 3-way valve to column 2.

Start immediately column 2 (19-mer synthesis) having completed cycle 1 nucleotide injection. Column drying: use the same procedure as for column 1 except use only 0.6 mL of 10% phenylisocyanate in pyridine injecting 0.2 mL immediately and then 50 µL at 1 min intervals. Leave reaction a total of 10 min and then begin Wash.

Deprotection and wash, cycle 1: procedure is the same as for column 1. After final pyridine wash select stop position.

Nucleotide addition, cycle 1: dissolve the first dimer (CT) in 0.35 mL pyridine (1 x reaction volume) by injecting pyridine into the flask. Transfer the solution into the MSNT vial (60 mg) by injection and when dissolved inject this on to the column. Inject 0.2 mL initially and then 50 µL every 1 min until all is injected. Use the same injection procedure as before. Wrap the column with aluminium foil and start counting the 60 min reaction time from the point of initial injection. Switch 3-way valve back to column 1. Cycles 2-9 (columns 1 and 2).

Further wash and nucleotide additions are carried out as described above. Use nucleotides as follows:

Cycle	Column 1	Column 2
2	C	TC
3	T	TT
4	A	TG
5	T	GC
6	A	TA
7	G	CA
8	C	GT
9	C	TG

On day 2 4-6 cycles should be carried out, the remainder being carried out on day 3. At the end of the last cycle carried out on day 2 wash the column with pyridine to remove nucleotide material and then chloroform, put the selector in the stop position and wrap the column with aluminium foil. Continue next day with the 10% TCA wash.

After assembly - columns 1 and 2: wash the column with pyridine for 10 min and then chloroform for 5 min. Transfer the contents of the column into a small sintered glass funnel using ether to wash out column. Wash beads several times with ether and suck dry briefly. The resin is now ready for deprotection. There should now be four samples of support (two complete syntheses and two intermediate samples).

4.2.6 Deprotection

Put each support from experiment 4.2.5 into a 2 mL plastic Eppendorf tube. Prepare a solution of syn-2-nitrobenzaldoxime (560 mg) in 8 mL of dioxane:water (1:1, v/v) and add 400 µL of 1,1,3,3-tetramethylguanidine. Put 2 mL of this orange solution into each Eppendorf tube and seal well with parafilm. Shake briefly and leave to stand overnight.

Filter the mixtures through small sintered glass funnels into 50 mL round bottomed flasks. Evaporate to dryness and dissolve the residue in 5 mL concentrated ammonia solution. Seal the flask well - use grease around joint. Incubate at 60°C for 3 h or 50°C for 5 h. Cool flask in ice and open carefully, remove vacuum grease from joint with hexane, and evaporate carefully to dryness (water pump).

Dissolve the residue in 5 mL of acetic acid:water (8:2, v/v), leave at room temperature for 30 min and evaporate to dryness. Take up in 3 mL of water and extract with 10 mL ether 5 times, each time shaking the flask vigorously for 30 sec, allowing the layers to separate and removing the top (ether) layer with a pipette. Evaporate again to dryness and take the residue up in water (2 mL) ready for purification.

4.2.7 hplc purification

4.2.7.1 Ion exchange chromatography

Set up a gradient hplc system with a Partisil-10 Sax column and buffers of 1 mM and 0.3 M KH_2PO_4, pH 6.3, in formamide:water (6:4, v/v) as "A" and "B" solvent inlets of pumps respectively. A new column should be washed with high strength buffer (B) for 30 min and then pre-equilibrated with low strength (A) until the recorder baseline is steady. Inject 1 µL - 2 µL of a standard of thymidine-5'-phosphate (1.2 mg in 300 µL water) with the detector on 0.2 A_{270} full scale deflection. Run a gradient of 0-10% buffer B over 45 min. One sharp peak should be seen indicating a good quality column.

Re-equilibrate the column for at least 10 min with buffer A and inject about 20-50 µL of oligonucleotide sample. Run a gradient of 0-50% B over 45 min for short chains or 0-70% B for long chains and observe chromatogram. Further runs can now be carried out changing the gradient if necessary to give maximum resolution. Collect material eluting at the column outlet when the desired peak is seen on the chromatogram. Do as many runs of 10-mer in particular as time allows and at least one run of 19-mer and measure the total amount of A_{260} in each peak. Note that to properly resolve the 10-mer a column temperature of 40°C is necessary, otherwise the product peak may be broad or even multiple due to self-association of chains.

Evaporate to dryness the eluates from the ion exchange column and take up in 0.1 volume of ethanol:water (2:8, v/v). Apply to a Biogel P2 column (100 cm x 2.5 cm) made up and prewashed in ethanol:water. Collect fractions and pool the material in the peak eluting at the void volume. A uv-monitor is useful but if not available measure the absorbance of individual fractions and plot a graph. Evaporate to dryness and take up in an appropriate volume of water to give a concentration of 2-5 A_{260}-units·mL^{-1}.

4.2.7.2 Reversed phase chromatography

Set up the gradient hplc with a µ-Bondapak C18 column (or equivalent) and 0.1 M ammonium acetate solution as buffer A and 0.1 M ammonium acetate:acetronitrile (2:8, v/v) as buffer B. Check the column using a standard of thymidine-5'-phosphate (1.2 mg in 400 µL) injecting about 1-2 µL on a scale of 0.2 A_{270}. Run a gradient of 0-15% B over 45 min.

The elution position of oligonucleotides is very sensitive to slight changes in concentration of acetonitrile. For a first run try the following conditions: 8% B for 4 min and then 8-15% B over 45 min. Warm samples to 50°C before injection and apply about 20-50 µL on a scale of 0.2 A_{270}. Observe the chromatogram and adjust the gradient if necessary. If the chromatogram shows 95% or greater in one peak then preparative purification is not normally necessary. Otherwise inject on a preparative scale and collect the material in the desired peak. The buffer is volatile and oligonucleotides can be obtained merely by evaporation or lyophilisation. Do as many column runs as time allows.

Estimation of yields: calculate yields after ion exchange and then again after reversed phase if this is carried out preparatively.

overall % yield =

$$\frac{A_{260}\text{ obtained}\cdot 100}{\text{fraction of material injected on column}\cdot \Sigma\varepsilon_{260}\text{ of each monomer}\cdot \mu\text{mol of starting resin}}$$

Notes:

(1) ε_{260} T = 8.8, C = 7.3, G = 11.7 A = 15.4 $[cm^2 \cdot \mu mol^{-1}]$

(2) No allowance is made for hypochromicity of oligonucleotide and therefore actual yields may be a little higher in some cases.

4.2.7.3 Purification by acrylamide gel electrophoresis (page)

Set up a Biogel P2 column and desalt separately about half of the products from experiment 4.2.6 (10-mer and 19-mer mixes). The method is as described for experiment 4.2.7 above.

Measure the absorbance at 260 nm of both products (eluted at the void volume). Evaporate to dryness and make up solutions in water at about 0.3 A_{260} units$\cdot\mu L^{-1}$.

Gel electrophoresis: the full procedure for isolation of oligonucleotides from gel slices is given in Appendix 4. Also given in Appendix 4 are the gel recipes and buffers needed below.

Make up a 20% acrylamide urea gel solution (see Appendix 4) (50 mL for a 0.35 mm gel or 100 mL for a 1.5 mm gel) in 1 M Tris-borate, pH 8.3, 0.1 M EDTA (TBE) buffer. Prepare a gel sandwich using two pre-washed 20 cm x 40 cm glass gel plates and side spacers well taped around. (The plate with the cut away section is usually siliconised to allow the plate to be removed after running, leaving the gel adhered to the other plate.) Add ammonium persulphate and tetramethylethylene diamine (TEMED) and pour gel being careful not to include air bubbles. Insert a comb with at least 1 cm slots and allow to set 1-2 h before running. When ready set up ready for loading.

On a thin gel (0.35 mm) only about 5 µL can be loaded per 1 cm slot width. On a thick gel (1.5 mm) this increases to about 20 µL per 1 cm width.

With samples in small Eppendorf tubes add one third of their volume of formamide loading dye-buffer mix. Boil for 2 min on a water bath and load samples. Make sure slots are carefully washed out with TBE running buffer before loading.

Run the gel at 1.5 kV, 11 mA for thin gels or 0.9 kV, 30 mA for thick gels until the fast blue dye reaches at least two thirds down the plate. Disconnect gel and remove gel plates transferring gel carefully on to Saranwrap. Wrap gel completely in Saranwrap and place on the plate containing a fluorescent indicator. Observe bands under a uv-lamp in the dark room and make an assessment of the success of the synthesis by comparing the mobilities with the known standards provided.

This whole procedure can be done after ^{32}P-labelling - see appendix 4 - and the eluted band sequenced by either Sanger-Brownlee (described) or Gilbert-Maxam procedures. Remember, however, that the labelled products migrate differently to the unlabelled oligonucleotides and that the eluted band here will probably not be chemically homogeneous.

REFERENCES

[1] Merrifield, R.B. (1963) J. Amer. Chem. Soc. 85, 2149
Letsinger, R.L. and Kornet, M.J. (1963) J. Amer. Chem. Soc. 85, 3045

[2] Gait, M.J. (1980) in "Polymer-supported reactions in organic synthesis", ed. P. Hodge and D.C. Sherrington, John Wiley, p 435;
Mathur, N.K., Narang, C.K., and Williams, R.E. (1980) in "Polymers as aids in organic synthesis", Academic Press, p 81

[3] Gait, M.J. and Sheppard, R.C. (1977) Nucleic Acids Research 4, 1135 and 4391; idem (1979) ibid 6, 1259

[4] Narang, C.K., Brunfeldt, K., and Norris, K.E. (1977) Tetrahedron Letters, 1819

[5] Miyoshi, K. and Itakura, K. (1980) Nucleic Acids Research Symposium Series No. 7, 281

[6] Miyoshi, K. and Itakura, K. (1979) Tetrahedron Letters, 3635;
Miyoshi, K. and Itakura, K. (1980) Nucleic Acids Research 8, 5491

[7] Gait, M.J., Singh, M., Sheppard, R.C., Edge, M.D., Greene, A.R., Heathcliffe, G.R., Atkinson, T.C., Newton, C.R., and Markham, A.F. (1980) Nucleic Acids Research 8, 1081

[8] Gait, M.J., Popov, S.G., Singh, M., and Titmas, R.C. (1980) Nucleic Acids Research Symposium No. 7, 243

[9] Duckworth, M.L., Gait, M.J., Goelet, P., Hong, G.F., Singh, M., and Titmas, R.C. (1981) Nucleic Acids Research 9, 1691

[10] Gough, G.R., Brunden, M.J., and Gilham, P.T. (1981) Tetrahedron Letters 22, 4177

[11] Matteucci, M.D. and Caruthers, M.H. (1980) Tetrahedron Letters 21, 719

[12] Atherton, E., Brown, E., Sheppard, R.C., and Rosevear, A. (1981) J. Chem. Soc. Chem. Commun. 1151

[13] Gait, M.J., Matthes, H.W.D., Singh, M., and Titmas, R.C. (1982) J. Chem. Soc. Chem. Commun. 37

[14] Chattopadhyaya, J.B. and Reese, C.B. (1978) J. Chem. Soc. Chem. Commun. 639

[15] Kierzek, R., Ito, H., Bhatt, R., and Itakura, K. (1981) Tetrahedron Letters 22, 3761

[16] Matteucci, M.D. and Caruthers, M.H. (1981) J. Amer. Chem. Soc. 103, 3185

[17] Itakura, K., personal communication

[18] Gait, M.J. and Matthes, H., unpublished results

[19] Reese, C.B., Titmas, R.C., and Yau, L. (1978) Tetrahedron Letters 2727

[20] Reese, C.B. and Zard, L. (1981) Nucleic Acids Research 9, 4611

ACKNOWLEDGEMENTS

We should like to thank sincerely Bob Sheppard for his continued encouragement and support and Joan Illsley for typing the manuscript. H.W.D.M. is supported by an EMBO Fellowship.

The following companies generously contributed specialist equipment or materials to this course:

Omnifit, Ltd.	- bench synthesiser
LKB Biochrom	- polymer support
Cruachan Chemical Co.	- nucleotide monomers, coupling agent and deprotecting agent

Appendix 1: Equipment list

1. Oligonucleotide assembly

	Supplier
1. 1 bench synthesiser - dual column model (see attached parts list)	Omnifit, Ltd., 51 Norfolk Street, Cambridge CB1 2LE.
2. 1 cylinder of Argon with pressure gauge	
3. 1 connector from Argon cylinder to join to 3 mm Teflon tubing	
4. 2 chemically inert syringes with replaceable needles	
either - 1750 RN (81230) 0.5 mL + 89728 22 spare needles (pack of three)	Hamilton or
or - 19925-1, 1 mL + 925-N2-22 spare needles (pack of two)	Glenco
5. 1 rotary evaporator with water bath, connected to high vacuum pump (not water pump). A useful addition is a liquid inlet to allow solvents to be delivered to the rotating flask without need to remove flask from vacuum.	Buchi
6. 20 small ground joint pear shaped flasks (5 mL)	
7. 20 Silicone rubber septum/stoppers "Subaseal" size no. 30 (to fit flasks) SYJ-350-200T (four packs of five)	Gallenkamp, (Gebrüder Haake GmbH in Germany)
8. 2 boxes of 12 x 1 mL microvials	
either - MV 10	Camlab, Ltd. Nuffield Road, Cambridge CB4 1TH. or
or - 13221	Pierce Chemical Co., 44 Upper Northgate Street, Chester, Cheshire CH1 4EF.
9. 24 Teflon/silicone discs to fit microvials (e.g.) 12712	Pierce Chemical Co.
4 stainless steel hplc solvent prefilters	Jones Chromatography or any hplc supplier
1 Teflon stirring bar	
1 magnetic stirrer	
tool kit: scalpel, pliers, screwdriver	
2 stop watches	

2. Functionalisation of resin

1. 1 - 15-20 mL glass reaction vessel with ground joint glass stopper at top and sintered glass disc and tap at bottom (see fig. 5)
2. 2 x 25 mL round bottomed flasks with glass stoppers
3. 1 stand and clamp for reaction vessel
4. 5 sintered glass funnels with small joint connector and side arm, FDK 360G — Gallenkamp
5. spatula (small)
6. 5 - 2 mL glass tubes or bottles
7. 1 oven at about 100°C

3. Deprotection

1. 1 water bath at 60°C
2. a few 2 mL Eppendorf tubes No. 3812
3. a few Pasteur pipettes, glass
4. PVC plastic tape

4. hplc purification

1. 1 hplc capable of programmed gradient elution with uv-monitor preferably with wavelength at 270-280 nm, injector with 1 mL loop, and appropriate stainless steel connectors for column
2. 1 syringe for loading of 100 µL samples
3. 1 syringe for loading of 500 µL samples
4. 1 analytical column Partisil-10 SAX, 25 cm x 0.4 cm — e.g. Whatman
5. 1 analytical column µ-Bondapak C18, 25 cm x 0.4 cm — Waters

5. Gel purification

1. 1 high voltage power pack — Raven Scientific
2. 1 - 20 cm x 40 cm gel kit with 1 mm spacers and 2 cm slots
3. yellow tape for making gel sandwich
4. 1 - uv-lamp 254 nm (in dark room)
5. kieselgel F254 tlc places (for visualisation of bands) — e.g. Merck
6. Saranwrap

6. Desalting

   1. 1 glass column, 100 cm x 2.5 cm (for Biogel P2) operated by gravity flow or peristaltic pump
   2. 1 - uv-monitor/spectrophotometer (260-280 nm) either in line with column or separate

7. General solvents and liquid reagents

		procedure
1.	pyridine	1 L redistilled from KOH, stored over molecular sieve 4Å
2.	N,N-dimethylformamide	1 L redistilled *in vacuo*, stored over molecular sieve 4Å
3.	chloroform	1 L freshly passed through basic Alumina (Woelm, activity grade Super I), stored over molecular sieve 4Å
4.	dioxane	50 mL as above
5.	dichloromethane	1 L as above
6.	piperidine	50 mL redistilled *in vacuo* under N_2
7.	concentrated ammonia solution	20 mL
8.	glacial acetic acid	20 mL
9.	ethyl alcohol 95%	2 L
10.	diethyl ether, anhydrous	50 mL
11.	phenyl isocyanate	1 mL AR grade, preferably distilled

8. Resin functionalisation

   1. dicyclohexylcarbodiimide, 10 g, known good quality or redistilled *in vacuo* under N_2
   2. 5'-O-dimethoxytrityl-2'-deoxynucleoside-3'-O-succinate (Appendix 2)
   3. ninhydrin reagents in small dropping bottles (10 mL of each):

      (1) 2 mL of 1 mM KCN diluted to 100 mL with pyridine
      (2) 500 mg of ninhydrin in 10 mL ethanol
      (3) 80 g phenol in 20 ml ethanol

9. Oligonucleotide assembly

		suppliers
1.	trichloroacetic acid, anhydrous - a fresh unopened bottle, 100 g	
2.	four protected mononucleotides (appendix 3)	Collaborative, Cruachan, P.L., Biosearch
3.	appropriate dinucleotide blocks (appendix 3)	Fluka, P.L. Biosearch
4.	mesitylenesulphonyl-3-nitro-1,2,4,triazole (appendix 3)	Cruachan, Biosearch
5.	molecular sieve beads type 4Å	

10. Deprotection

1.	syn-2-nitrobenzaldoxime, 10 g	Cruachan
2.	1,1,3,3-tetramethylguanidine, puriss, 20 mL	

11. Gel purification

1. urea, specially purified, 500 g
2. acrylamide, specially purified, 100 g
3. ethylene bis-acrylamide, specially purified, 10 g
4. amberlite MB1, mixed bed ion exchange resin, 50 g
5. 5 x TBE buffer (0.5 M Tris borate, pH 8.3, containing 10 mM EDTA), 1 L
6. ammonium persulphate, 20 g
7. tetramethylethylenediamine, puriss, 5 mL
8. gel loading solution: 8 mL formamide
100 μL 0.5 M EDTA
2 mg bromophenol blue
5 mg xylene cyanol FF
made up to 10 mL with water
9. Biogel P2 (50 - 100 mesh) — Bio-Rad

12. hplc purification

1.	potassium dihydrogen orthophosphate AR a 1 M solution in water, pH 6.3	1 L
2.	ammonium acetate - a 1 M solution in water, pH unadjusted	1 L
3.	acetonitrile, hplc grade	2 L
4.	formamide, AR or hplc grade	2 L

13. Miscellaneous (chemicals used in appendix 2 and 3)

1.	succinic anhydride, AR	B.D.H.
2.	4-dimethylaminopyridine	Aldrich
3.	fluorenylmethoxycarbonylglycine (FmocGly)	Serva or Fluka
4.	Dowex 50-x8 ion exchange resin AR grade (H+ form). Prewash with 1 M NaOH, water, 1 M HCl, water, pyridine:water (2:8, v/v) then store at 4°C	

14. Parts list for bench synthesiser (Dual column)

Supplier: Omnifit, Ltd., 51 Norfolk Street, Cambridge CB1 2LE
Tel. (0223) 69841, Telex 817135

Quantity	Cat. No.	Description
1	3102	pressure stat, 50 psi ≙ 3.5 bar
4	3200	3-valve inert reservoir, 1 L
1	3210	3-valve inert reservoir, 100 mL
2	6620	chromatography tube, 50 mm
2	6641	septum injector endpiece
2	6643	variable length endpiece
1	3600	mounting for valve and column (includes: 4 column clips, nut and bolt, Rheodyne 6-way valve and 2-way valve)
4	2320	plugs
1	1102	3-way valve
1	1006	8-way connector
1	6212	25 x 150 mm complete column (for silica gel drying tube)
1	2301	2-way coupling, polypropylene
2	2110	tube end fitting, 1/16" mixed
2	2210	tube end fitting, 1/8" mixed
2	2312	gripper fitting, 1/8"
2	3011	teflon tubing, 1.5 mm, i.d.
2 m	3001	teflon tubing, 0.8 mm, i.d.
2 m	3005	teflon tubing, 0.3 mm, i.d.
1		special spare parts kit:
10		006 silicone O-rings
10		teflon coated septum
10		006 fluoron O-rings
10	6552	frits (teflon)
2	2310	gripper fitting, 1/16"
4	6654	brass pinch clamp

Appendix 2: Functionalisation of support[1)]

Volumes: ca 5 mL per gram of dry resin[2)]

Treatment[3)]	No. of	Time per wash [min]	Ninhydrin test[5)]
anhydrous ethylene diamine	1	16 h	
DMF	10-15	2	resin +, last eluant -
10% diisopropylethylamine: DMF (v/v)	3	5	
DMF	5	2	
(Fmoc GlyO)$_2$:DMF[4)]	1	90	
DMF	10	2	resin -
piperidine:DMF (2:8, v/v)	1	10	
DMF	5	2	resin +
(FmocGlyO)$_2$=:DMF[4)]	1	90	
DMF	10	2	resin -

The resin may now be washed with CH_2Cl_2, dioxane, ether, dried and stored.

or continue:

piperidine:DMF (2:8, v/v)	1	10	
DMF	5	2	resin +
nucleoside succinate[6)] anhydride	1	90	
DMF	5	2	resin -
CH_2Cl_2	5	2	
dioxane	5	2	
ether	5	2	

The resin may now be dried and stored[7)].

Notes:

[1)] This should be carried out in a glass vessel fitted with a ground glass joint and stopper. On the bottom should be a sintered glass frit and stopcock. Filtration is by slight nitrogen pressure to the top of the vessel (see Fig. 5, p. 42).

[2)] Addition of solvents and reagents is batchwise through the ground glass joint.

[3)] Resin should be gently agitated by shaking for the first 0.5 to 1 min of each wash and allowed to stand for the remaining wash time.

[4)] Preparation of FmocGly anhydride and reaction with resin: Ten equivalents of FmocGly (over resin loading of sarcosine) is dissolved in CH_2Cl_2 with the minimum quantity of DMF added to obtain solubility. Dicyclohexyl-carbodiimide (DCC) (5 equiv) dissolved in dichloromethane is added. After 15 min the mixture is evaporated to dryness, dissolved in 5 mL DMF and filtered through a small sintered funnel. The filtrate is transferred immediately to the resin.

5) A small sample of resin is washed on a sinter with DMF, dichloromethane and diethyl ether and a ninhydrin test carried out (E. Kaiser et al. (1970) Anal. Biochem. 34, 595-598) on the beads. The test can also be carried out on one drop of eluant.

6) Preparation of nucleoside succinate anhydride and reaction with the resin: Ten equivalents of nucleoside succinate is dissolved in dichloromethane. DCC (5 equiv) is added. The reaction and work up is exactly as described for FmocGly anhydride.

7) Final dried resin is tested for glycine by overnight hydrolysis in a sealed tube with 6 N HCl, and assayed using an amino acid analyser with α-amino-β-guanidine propionic acid as internal standard. Trityl is estimated spectrophotometrically in 60% perchloric acid:ethanol (3:2, v/v) $\varepsilon_{495} = 71.7\ cm^2 \cdot \mu mol^{-1}$

1. General procedure for preparation of pyridinium 5'-O-dimethoxytrityl-2'-deoxynucleoside-3'-O-succinates

The 5'-O-dimethoxytrityl deoxynucleoside (dT, dbzA, dbzC or dibG) (3.1 mmol) is dissolved in pyridine (6.25 mL) and dimethylaminopyridine (0.3 g) added followed by succinic anhydride (3.4 mmol, 0.34 g). The mixture is kept at room temperature for 12 h and then applied to a Dowex 50-X8 (pyridinium form) ion exchange column (2 cm x 10 cm). The column is eluted slowly with pyridine:water (1:4, v/v). The eluate is evaporated to dryness and the residue co-evaporated with toluene to a foam. This is dissolved in dichloromethane and applied to a short fat silica gel column (100 g, Kieselgel 60H, Merck 7736) eluting first with dichloromethane and then ethanol:dichloromethane (3:97, v/v) to elute the product. It is essential to pool fractions conservatively making sure no higher or lower R_f material is present. It is particularly important to ensure the absence of succinic acid which will seriously impair efficiency of subsequent reactions. This can occasionally be seen to be eluted later from the column as a trityl negative, low R_f spot. Column fractions should be monitored by tlc in ethanol:dichloromethane (10:90, v/v). Pure fractions are evaporated to dryness and precipitated into ether:pentane (3:2, v/v).

Appendix 3:

1. Preparation of DMTrdT

2'-Deoxythymidine (10 g, 41.2 mmol) was dissolved in anhydrous pyridine (100 mL) and the solution evaporated to dryness. The resultant solid was dissolved in 40 mL pyridine and 4,4'-dimethoxytritylchloride (15 g, 44.3 mmol) was added. The mixture was shaken in the dark for 4 h in a sealed flask, whereupon some crystals had separated. Silica gel tlc of the product in ethanol:chloroform (15:85, v/v) showed one trityl and uv-positive spot at higher R_f than starting material. 4 mL methanol was added, the mixture evaporated to a gum and partitioned between 200 mL chloroform and 80 mL cold 1 M $NaHCO_3$. The organic phase was washed with water (40 mL) and evaporated to dryness in the presence of pyridine. After co-evaporation with toluene the product was dissolved in dichloromethane-0.1% pyridine and chromatographed on Kieselgel 60H (400 g, 13 cm diameter) using 500 mL of dichloromethane-0.1% pyridine and 3 L of 5% ethanol:dichloromethane-0.1% pyridine. Pure fractions were pooled, evaporated to a foam with pyridine, dissolved in 80 mL chloroform and precipitated with pentane (3 L).

Yield: 15.9 g (71%)

2. Preparation of DMTrdbzA

To 2'-deoxyadenosine (10 g, 39.8 mmol) was added 200 mL dry pyridine and the suspension evaporated to dryness. This procedure was repeated and the resultant solid suspended in 240 mL pyridine. To the ice cooled suspension was added redistilled benzoyl chloride (20 mL) and the mixture shaken in the dark in a sealed flask for 2 h. Silica gel tlc in ethanol:chloroform (10:90, v/v) showed one high R_f spot. The reaction mixture was partitioned between cold 1 M $NaHCO_3$ (800 mL) and chloroform (800 mL) and the aqueous layer washed with chloroform (2 x 400 mL). The combined chloroform layers were washed with 400 mL 1 M $NaHCO_3$ and evaporated in the presence of pyridine to an oil. The oil was dissolved in 275 mL pyridine:methanol (36:8, v/v), cooled to -30°C and 47 mL 5 N NaOH added. The mixture was poured into excess Dowex pyridinium ion exchange resin and then filtered slowly through a column of Dowex using pyridine:water (1:4, v/v) as eluant. The eluate was washed with ether (6 x 600 mL) and evaporated to a gum with pyridine.

The resultant gum was dissolved in pyridine (100 mL) and 4-4'-dimethoxytrityl chloride (15 g, 44.3 mmol) added. The reaction mixture was evaporated slightly and left in the dark for 2 h in a sealed vessel. Tlc showed complete reaction. Methanol (40 mL) was added and the mixture evaporated to a mobile oil (N.B. do not leave on evaporator for a long period) and partitioned between 1 M $NaHCO_3$ (200 mL) and chloroform (500 mL). The aqueous layer was washed twice with 200 mL chloroform and the combined chloroform layers washed with water (2 x 200 mL) and evaporated to a gum in the presence of pyridine. The product was co-evaporated with toluene to a foam and chromatographed in two portions on Kieselgel 60H by the short column method (400 g, 13 cm diameter) eluting with 500 mL of dichloromethane-0.1% pyridine, 500 mL 2% ethanol:dichloromethane-0.1% pyridine and then 3 L of 5% ethanol:dichloromethane-0.1% pyridine. Pure fractions were pooled, evaporated to a foam with pyridine, dissolved in chloroform (50 mL) and precipitated into 2 L pentane:ether (2:1, v/v).

Yield: 24.5 g (93%)

3. Preparation of DMTrdibG

To 2'-deoxyguanosine (20 g, 74.9 mmol) was added 100 mL dry pyridine and the suspension evaporated to dryness. This procedure was repeated and the resultant solid suspended in 400 mL chloroform and 100 mL pyridine. To the ice-cooled suspension was added dropwise isobutyryl chloride (50 mL, 480 mmol) and the mixture shaken in the dark for 2 h. Silica gel tlc in ethanol:chloroform (10:90, v/v) showed one high Rf spot. Chloroform was added (400 mL) and the mixture washed with 800 mL cold 1 M $NaHCO_3$. The aqueous phase was backwashed with 200 mL chloroform and the combined chloroform layers washed again with 400 mL 1 M $NaHCO_3$. The aqueous phase was backwashed with 100 mL chloroform and the combined chloroform layers evaporated to an oil in the presence of pyridine. The oil was dissolved in 150 mL pyridine and 200 mL tetrahydrofuran, cooled to -30°C and 2 N NaOH (300 mL) carefully added. The mixture was swirled continuously for 20 min at 0°C when tlc showed complete reaction. The barely clear solution was poured into excess Dowex pyridinium ion exchange resin using pyridine:water (1:4, v/v) as eluant. The eluate was washed with ether (4 x 700 mL) and evaporated to a gum with pyridine. The gum was dissolved in 400 mL anhydrous pyridine and 4,4'-dimethoxytrityl chloride (33 g, 97.5 mmol) added. The reaction mixture was left in the dark for 2 h in a sealed vessel. Tlc showed complete reaction. Methanol (80 mL) was added and the mixture evaporated to a mobile oil and products partitioned between chloroform (800 mL) and 1 M $NaHCO_3$ (250 mL). The aqueous layer was washed with 200 mL chloroform and the combined chloroform layers washed with water (2 x 250 mL) and evaporated to a gum in the presence of pyridine and finally to a foam with toluene. The product was chromatographed in three portions on Kieselgel 60H by the short column method (400 g, 13 mm diameter) eluting with 500 mL dichloromethane-

0.1% pyridine, 500 mL of 2% and 3 L of 7% ethanol:dichloromethane-0.1% pyridine. Pure fractions were pooled, evaporated to a foam with pyridine, dissolved in chloroform and precipitated into pentane:ether (2:1, v/v).

Yield: 35.1 g (73%)

4. Preparation of DMTrdbzC

2'-Deoxycytidine hydrochloride (10.55 g, 40 mmol) was dried by evaporation of pyridine (2 x 150 mL) and then suspended in 300 mL dry pyridine. Diisopropyl ethylamine (7.05 mL, 40.5 mmol) was added and after 5 min pentachlorophenyl benzoate (74.09 g, 200 mmol) was added. The mixture was stirred at 85°C under dry nitrogen for 15 h when silica gel tlc in EtOH:$CHCl_3$:pyridine (10:90:0.1, v/v) showed the benzoylation to be complete (intense uv-positive spot of R_f ∿0.2 due to dbzC). The intermediate product was not isolated but tritylated directly: 4,4'-dimethoxytrityl chloride (14.9 g, 44 mmol) was added and the orange mixture was left sealed for 2 h at room temperature, when tlc showed reaction to be complete (intense uv and trityl positive spot of R_f ~0.39 due to the desired DMTrdbzC). Methanol (40 mL) was added and solvent was removed <u>in vacuo</u> to leave an oil plus some crystalline material. Chloroform (500 mL) was added and the mixture was filtered to remove the crystalline, excess, pentachlorophenyl benzoate. The filtrate was washed with cold 1 M $NaHCO_3$ (150 mL) and the aqueous phase was backwashed with chloroform (2 x 150 mL). The combined organic phases were then washed with water (2 x 300 mL) and the clear orange solution was evaporated <u>in vacuo</u> in the presence of pyridine to a viscous oil plus more crystalline pentachlorophenyl benzoate. Residual pyridine was removed by co-evaporation with toluene, and the residue was dissolved in dichloromethane-0.1% pyridine (150 mL). Some 30 g of excess white crystalline pentachlorophenyl benzoate were recovered at this point by filtration. The filtrate was chromatographed on Kieselgel 60H (800 g silica, 13 cm diameter column) by the short column method using 1 L dichloromethane-0.1% pyridine, 2 L of 5% and finally 4 L of 6% ethanol in dichloromethane-0.1% pyridine as eluant. Pure fractions were combined, evaporated to a foam in the presence of pyridine, the foam was dissolved in dichloromethane-0.1% pyridine and the product precipitated by dropwise addition of the solution to vigorously stirred pentane:ether (1 L, 2:1, v/v).

Yield: 70-80%

5. General procedure for preparation of triethylammonium salts of protected deoxynucleoside 3'-O-(2-chlorophenyl phosphates)

A mixture of 1,2,4-triazole (2.76 g, 40 mmol; recrystallised from anhydrous dioxan), dry triethylamine (4.87 mL, 35 mmol), 2-chlorophenyl phosphorodichloridate (3 mL, 15 mmol), and dry THF (100 mL; distilled from sodium wire and benzophenone) was stirred for 30 min at room temperature with exclusion of moisture (a copious white precipitate of triethylamine hydrochloride formed immediately). The mixture containing 2-chlorophenyl phosphorodi(triazolide) was filtered in a dry box into a pre-dried (by evaporation of dry THF containing 1% triethylamine) protected deoxynucleoside (10 mmol; viz. 5.45 g of DMTrdT, 6.34 g of DMTrdbzC, 6.58 g of DMTrdbzA, or 6.40 g of DMTrdibG) and the clear pale yellow solution was left sealed for 1 h at room temperature. Silica gel tlc in ethanol:chloroform (1:9, v/v) showed reaction to be complete with a trityl positive spot near the baseline. Added 1 M triethylammonium bicarbonate (100 mL, pH 8) and the mixture was then extracted with 300 mL chloroform. The organic phase was washed with water (3 x 150 mL), any emulsions formed were separated by centrifugation, and evaporated _in vacuo_ in the presence of pyridine to a foam. Material to be used for the synthesis of fully protected dimers was obtained by precipitation from pentane:1% triethylamine by dropwise addition of a dichloromethane solution. The precipitate was collected by centrifugation, washed twice with dry ether and then dried _in vacuo_ over P_2O_5.

Yields: generally 85-95%

N.B. Material to be used for synthesis by monomer addition was subjected to purification by short column chromatography on silica gel 60H (200 g silica, 7 cm diameter) using 2 L of 10% ethanol in dichloromethane:1% triethylamine as eluant. Pure fractions were pooled and evaporated _in vacuo_ in the presence of pyridine to a glass. The pure product was then precipitated as above.

6. General procedure for preparation of fully protected dimers (1.5 mmol scale)

6.1 Preparation of the hydroxyl component

The triethylammonium 5'-O-DMTr-deoxynucleoside-3'-O-2-chlorophenylphosphate (6.6 mmol) was co-evaporated three times with pyridine to give a final volume of 65 mL. 3-hydroxypropionitrile (Aldrich, stored over molecular sieve 4Å) (2.2 mL, 32.2 mmol) was added followed by mesitylenesulphonyl-3-nitro-1,2,4-triazolide (MSNT, 3.85 g, 13 mmol). After 1 h at room temperature silica gel tlc in ethanol:chloroform

(1:9, v/v) showed almost complete conversion to a spot R_f 0.5-0.7. Ethyl acetate (450 mL) was added and the mixture washed with 0.1 M $NaHCO_3$ solution (3 x 150 mL) and then saturated sodium chloride solution (150 mL). The organic phase (and any necessary backwashings) was evaporated to an oil with pyridine and then co-evaporated twice with toluene. The product was dissolved in 50 mL chloroform and cooled in ice. A precooled solution of trichloroacetic acid (3.19 g, 19.5 mmol) in 50 mL chloroform was added and the dark orange solution left at 0°C (the reaction is followed by tlc in 10% ethanol:chloroform and is usually complete within 10 min). Pyridine (10 mL) was added and the solution diluted with 400 mL chloroform and washed with 0.1 M $NaHCO_3$ solution (3 x 150 mL) followed by saturated sodium chloride solution (150 mL). The organic phase (and any necessary backwashings) was evaporated to an oil in the presence of pyridine and divided into four portions as pyridine solutions.

6.2 Preparation of the dimer

To one of the four portions of 'hydroxyl component' (ca. 1.5 mmol) was added the appropriate triethylammonium 5'-O-DMTr-deoxynucleoside-3'-O-2-chlorophenylphosphate (2 mmol) and the mixture co-evaporated three times with pyridine to give a final volume of 20 mL. MSNT (1.19 g, 4 mmol) was added and the solution left at room temperature for 1 h. Silica gel tlc in 10% ethanol:chloroform (1:9, v/v) showed complete conversion of the hydroxyl component to higher R_f spot (or spots). Water (10 mL) was added and after 20-30 min chloroform (200 mL) was added and the mixture washed with 0.1 M $NaHCO_3$ solution (3 x 150 mL) followed by saturated sodium chloride solution (150 mL). On occasion emulsions were formed which were separated by centrifugation. The organic phase (and any necessary backwashings) was evaporated to an oil with pyridine, diluted with chloroform to ca. 15 mL and chromatographed on Merck 7736 Kieselgel 60H (45 g) by the short column method (diameter 7 cm). The column was eluted with 300 mL of dichloromethane-0.1% pyridine followed by a 1-1.5 L of ethanol in dichloromethane-0.1% pyridine (for composition see below). The eluate was monitored by tlc and pure fractions pooled, evaporated to an oil with pyridine, dissolved in chloroform and product precipitated with ether:pentane (1:2, v/v).

Yields: 54-85% (based on hydroxyl component)

% ethanol in dichloromethane	4	5	6	7
	TT	TC AC	TG CG	AG
	TA	CC AA	GT GC	GG
		CA AT	GA	
		CT		

6.3 Removal of cyanoethyl groups from dimers

The dimer (100 μmol) was dissolved in 2 mL anhydrous triethylamine: acetonitrile (1:1, v/v) in a 10 mL flask. The reaction was followed by tlc in ethanol:chloroform (1:9, v/v) and is usually complete within 1 h to give one spot on the baseline. The solution was evaporated to an oil under high vacuum, dissolved in 2-3 mL chloroform:0.1% pyridine and product precipitated by dropwise addition to 200 mL anhydrous ether vigorously stirred in a centrifuge tube. The precipitate was separated by centrifugation and decantation, washed twice with 100 mL ether and dried *in vacuo*.

Yields: 90-95%

6.4 Preparation of mesitylenesulphonyl-3-nitro-1,2,4-triazole

6.4.1 Preparation of 3-nitro-1,2,4-triazole

3-Amino-1,2,4-triazole (Aldrich, 50 g, 0.6 mol) and sodium nitrite (200 g, 2.9 mol) in water (300 mL) in a 2 L flask were stirred and cooled in ice:salt. Concentrated nitric acid (170 mL) was added dropwise maintaining the temperature at 0-10°C (N.B. considerable heat is evolved and frothing takes place). The mixture was allowed to warm to room temperature overnight and cooled once more to 0°C. The precipitate was collected by filtration at the pump, washed with a little iced water and dried *in vacuo* over P_2O_5. Recrystallisation from ethyl acetate yielded pale yellow plates (39.7 g, 58%). R_f=0.5 on silica gel tlc in ethyl acetate.

6.4.2 Preparation of mesitylenesulphonyl-3-nitro-1,2,4-triazole

Mesitylenesulphonyl chloride (recrystallised from n-pentane) (17.4 g, 79.6 mmol) and finely powdered 3-nitro-1,2,4-triazole (9.2 g, 80.0 mmol) were suspended in dry dioxan (230 mL) which was stirred and cooled in ice. Triethylamine (redistilled from $NaBH_4$) (11.1 mL) was added and the mixture stirred at room temperature for 1 h. Triethylammonium chloride was removed by filtration and the filtrate evaporated to a solid, which was recrystallised from anhydrous dioxan (50 mL).

Yield: 23 g (98%)   mp 135.5-137°C.

6.5 Preparation of pentachlorophenyl benzoate

Pentachlorophenol (133.2 g, 0.5 mol) was dissolved in dry pyridine (500 mL) and to the solution was added freshly distilled benzoyl chloride (63.8 mL, 0.55 mol) over a 10 min period, with stirring and exclusion of moisture (a crystalline precipitate formed during the addition). The mixture was then refluxed for 1 h, cooled down (a considerable quantity of the ester crystallised out at this point), and the bulk of the pyridine was evaporated in vacuo. The residue was dissolved in chloroform (1.5 L) and washed with 1 M $NaHCO_3$ (1 L). The aqueous layer was backwashed twice with 100 mL chloroform and the combined organic phases were washed with 1 L water and then dried over anhydrous sodium sulphate. The orange/pink solution was evaporated in vacuo to leave a pale pink crystalline solid (~190 g) which was recrystallised from 450 mL chloroform.

Pentachlorophenyl benzoate (160.3 g, 87%) was obtained as large colourless prisms, m.p. 161-162.5°C.

Appendix 4: Polyacrylamide gel electrophoresis

1. Preparation of 20% acrylamide gel

Make a 20% acrylamide stock solution by dissolving 210 g urea, 100 g acrylamide (specially pure) and 2.5 g ethylene-bis-acrylamide (specially pure) in distilled water to make a volume of about 350 mL. Add about 20 g of Amberlite MB1 mono bed ion exchange resin, swirl for 5 min, filter the mixture and make up the solution to 450 mL.

To 45 mL acrylamide stock add
5 mL 1 M Tris-borate, pH 8.3 containing 100 mM EDTA (10 x TBE buffer)
400 µL 10% ammonium persulphate solution (fresh)
40 µL tetramethylethylene diamine (TEMED)

Pour gel rapidly using pipette or syringe and set for 1-2 h in an almost horizontal position. Run gel in 1 x TBE buffer.

Gel loading solution
8 mL formamide
100 µL 0.5 M EDTA
2 mg bromophenol blue
5 mg Xylene Cyanol FF
made up to 10 mL with water

2. Elution and desalting of oligonucleotides from gel slices

Cut out appropriate band (visualised under uv) using a scalpel and put the gel slice into a siliconised glass tube. Add enough water to totally cover the gel (300-800 µL) and leave overnight at 5-20°C. Carefully pipette off the liquid and wash the gel slice twice more leaving for 1 h for each wash. Apply the combined solutions to a small DEAE Sephadex A25 column (1 mL) made up in a 5 mL disposable plastic pipette cut to size. The column is prewashed in 1 M NaCl, 10 mM Tris-HCl, pH 7.4 and then equilibrated in 10 mM Tris-HCl, pH 7.4. After sample application the column is washed with 5-10 mL of low salt buffer and product eluted in high salt buffer (usually in the first 1-2 mL). The solution is diluted by 4 with water and applied to a Biogel P2 column (30 cm x 1.5 cm) and eluted in ethanol:water (2:8, v/v). The oligonucleotide emerges at the excluded volume.

3. ^{32}P-labelling of oligonucleotides

Dry down 20 µL γ-^{32}P-ATP (3 Ci·µmol^{-1}, 1 mCi·mL^{-1}) in a siliconised glass tube (6 mL), dissolve in 9.5 µL of the following mixture

1 µL 3 µM oligonucleotide (3 pmol); 2.5 µL 4X kinase buffer (40 mM $MgCl_2$; 40 mM dithiothreitol (DTT); 200 mM Tris-HCl, pH 7.6, containing 3% spermidine)
add 6 µL H_2O to obtain a total volume of 9.5 µL.

Flow sheet:
heat at 90°C for 2 min
incubate at 37°C for 5 min
add 0.5 µL (5 units) T4 polynucleotide kinase (EC 2.7.1.78)
incubate at 37°C for 20 min
add 8 µL of gel loading solution
heat at 90°C for 2 min
apply to a 0.35 mm thin gel (20 cm x 40 cm) in a 2 cm slot
run at 1.5 kV (11 mA) until dye (fast blue) reaches the lower third of the plate
dismantle, cover with Saranwrap
autoradiograph for 10 min
cut gel band out with scalpel and put into siliconised tube
soak in 200 µL H_2O at 0°C overnight and collect liquid
soak twice more in 100 µL H_2O
check gel band by Cerenkov-counting before and after elution
to ensure efficiency (80-90% should be eluted)
difference is counts of radioactive oligonucleotide in solution.

Appendix 5: Wandering spot sequencing (See ,,Determination of sequences in RNA", G.G. Brownlee, North-Holland, Amsterdam, 1972)

1. Materials

Electrophoresis tank and 5 kV supply with buffer of 10% acetic acid: 7 M urea adjusted with pyridine to pH 3.5.

Cellulose acetate strips (Schleicher & Schüll); large sheet of Whatman

3MM paper or blotter; glass rods 3 mm x 30 cm, 2 per sequencing experiment, plus 2 extra; glass plates 20 cm x 40 cm, 1 per sequencing experiment; 3MM paper strips 2.5 cm x 35 cm, 6 per plate, plus 1 extra.

Polygram Cel.300 DEAE/HR-2/15 80.1 mm plastic plates 20 cm x 40 cm (Machery Nagel); medical wipes; double-sided Sellotape; 3MM paper 20 cm x 10 cm folded to 4 cm x 10 cm as wick, 1 per plate; No. 2 Terry clips, plastic coated, 4 per plate, water spray bottle (fine spray); 70°C oven, Petri dish; glass tank 20 cm x 20 cm with slots, 1 per 1 or 5 plates; glass tank 20 cm x 20 cm without slots, 1 per 4 or 5 plates, autoradiography equipment (cassettes, screens, old film, -70°C freezer, dark room, film developer, Sellotape, etc.); radioactive ink; polythene sheet stretched over a former;

Electrophoresis running buffer: (fresh)
186 mg EDTA
10 mL acetic acid
90 mL 7 M urea
adjusted to pH 3.5 with pyridine

snake venom phosphodiesterase (EC 3.1.4.1, SVPDE), 0.25 mg mL^{-1} in 50 mM Tris-HCl, pH 8.9, 10 mM $MgCl_2$; tRNA (30 mg$\cdot mL^{-1}$); ethanol. Dye marker solution: 1% Xylene Cyanol FF, 1% acid fuchsin and 1% methyl orange in water.

'Homomix' - an RNA digest. Make up several homomixes of different strengths of RNA and times of digestion. Pick the one which gives the best resolution for chains 1-20 long. Homomix may be re-used many times - just pour back into flask and store at 4°C.

Take 10 g of yeast RNA, dissolve in 100 mL of 1 M KOH and hydrolyse for 5 or 10 min at room temperature. Neutralise to pH 7.5 with concentrated HCl and dialyse against distilled water overnight. Check pH and readjust if necessary. Add 94.5 g urea and make up to 225 mL. This makes a 4% RNA digest. This may be diluted with 7 M urea to give a 3% or 2% mix as appropriate.

2. Calibration of SVPDE reactions

Rate of digestion is dependent on activity of enzyme, amount of junk in oligonucleotide, etc. Best results are obtained by carrying out a calibration experiment on each chain.

Take ca. 10,000-20,000 cpm of oligonucleotide (should be no more than 10-20% of total counts). Spot onto polythene sheet which is stretched out onto a former. Add 0.5 µL of tRNA and dry down in a vacuum desiccator to give a small white residue. Dissolve each spot in 7 µL of SVPDE in a 10 µL capillary and at various time interval (e.g. 0, 2, 10, 30 min) spot 1-2 µL onto a DEAE plate at about 5 cm up the plate and dry the spots. Using Terry clips attach a paper wick to the top of the plate and run in homomix at 65°C. Spray start in the usual way (see below) until blue dye is about two-thirds up the plate. Dry and autoradiograph overnight in cassette at -70°C.

From the resultant pattern of spots pick about 4 new time points suitable to give all possible digestion products represented.

3. SVPDE reactions

Prepare another 10,000 cpm of oligonucleotide. Add 0.5 µL tRNA and dry down on polythene as before. Dissolve spots in 7 µL of SPVDE and at time intervals decided as above take 1-2 µL samples and freeze at -20°C in capillaries.

When all digestions are complete, spot each time point onto the origin of a prepared cellulose acetate strip (see below) and allow to dry each time between each spotting. Spot on also some undigested oligonucleotide. Carry out two-dimensional separation as described below and autoradiograph to see positions of spots and read off sequence (see E. Jay, R. Bambara, P. Padmanabhan, and R. Wu (1974) Nucleic Acids Research 1, 331).

4. Two-dimensional ionophoretic fractionation procedure

The digest is fractionated in the first dimension at pH 3.5 on strips of cellulose acetate by high voltage ionophoresis using the up-and-over tanks. Precut strips of cellulose acetate are available from Schleicher & Schüll in 3 cm x 55 cm long strips. 'Cellogel' cellulose acetate sheets from Colab. have also been used successfully. The cellulose acetate strip is first wetted with the pH 3.5 buffer and, to avoid the inclusion of air bubbles, this should be done from one side by flotation. One end of the strip is placed on the surface of the buffer contained in a Petri dish. As it absorbs the buffer the rest of the strip is wetted by passing it slowly across the surface of the liquid. Finally the strip is completely immersed in buffer to ensure it is completely wet. This urea-containing buffer is also used in the buffer compartments of the electrophoresis tank. The point of application, about 10 cm from one end of the strip, is blotted free of excess liquid and the digest is applied as a small spot and allowed to soak in. A dye mixture of Xylene Cyanol FF (blue), acid fuchsin (red) and methyl orange (yellow) is applied on each side of the digest. Care has to be taken to avoid the strip drying out while the sample is applied. This is done by having the ends covered with wet tissue paper. Excess buffer is then removed from the strip by blotting and it is rapidly dipped in 'white spirit' (Varsol) to prevent evaporation of the buffer while the strip is put into the ionophoresis tank. The origin is at one end near the negative electrode vessel. Ionophoresis is carried out at between 50 and 100 V/cm (usually 25-35 min at 5 kV).

Plates are prepared by taping the plastic sheets of DEAE onto glass plates of the same size using double-sided Sellotape. Pencil marks are carefully made on the edges of the plate 5 cm above the bottom, marking the place where the transfer is to be made. Nucleotides are blotted from the cellulose acetate onto the plates as follows: The strip is removed from the tank used for the first-dimensional run and monitored with a portable Geiger counter to find the position of the oligonucleotides - usually in the region between the red and blue markers.

Place the strip lengthwise along a glass rod so that the area to be transferred is contacting the glass (but facing upwards). Place three moist, but not too wet, strips of Whatman 3MM paper carefully over the long edges of cellulose acetate strip (three on each side) so that the strip is held down on the table and stretched over the glass rod. Make a 'sandwich' by placing the DEAE plate inverted so that the application line on the plate touches the cellulose acetate strip along the entire length of the rod. Make sure there is another glass rod positioned further down under the plate to keep it from touching the table too near the application area. Hold the sandwich in place by very carefully putting a weight of about 1 kilo on the top so that the plate presses down evenly onto the cellulose acetate strip. Water thus flows from the 3MM paper, transferring the oligonucleotides onto the DEAE-cellulose layer. It is preferable to use strips that still have some 'white spirit' on their surface, rather than let the buffer in the strip dry out and to allow 20-30 min for the transfer. In order to ensure efficient transfer, the paper strips are wetted with water occasionally. When the dye markers are transferred onto the plate the transfer is usually complete. Take the plate and dip it in a trough of ethanol to wash off urea and salts. Clip a 3MM wick to the top of the plate using Terry clips (this is not always necessary and depends how far the plate is developed).

Fractionation in the second dimension is carried out at 65°C in an incubator using one of the homomixes. A tall tank is formed by inverting a thin-layer 20 cm x 20 cm tank over another one (usually with slots to accept several plates). Make sure during chromatography that the joint between the two tanks is well greased to prevent evaporation of the solvent. The tank is equilibrated with about 100 mL of the homomixture at 60°C in the incubator for 1 h. The plate is also equilibrated at 60°C, but outside the tank for the same time. Before starting the ascending homochromatography, it is very important to spray the region of the origin with distilled water. After this, the plate is transferred into the equilibrated tank and ascending homochromatography continued until the front reaches the top of the plate. This usually takes about 5 h, although there is a considerable variation in running time.

Remove the plate, dry and apply radioactive ink in convenient locations to orient the plate. Autoradiograph in a cassette at -70°C overnight.

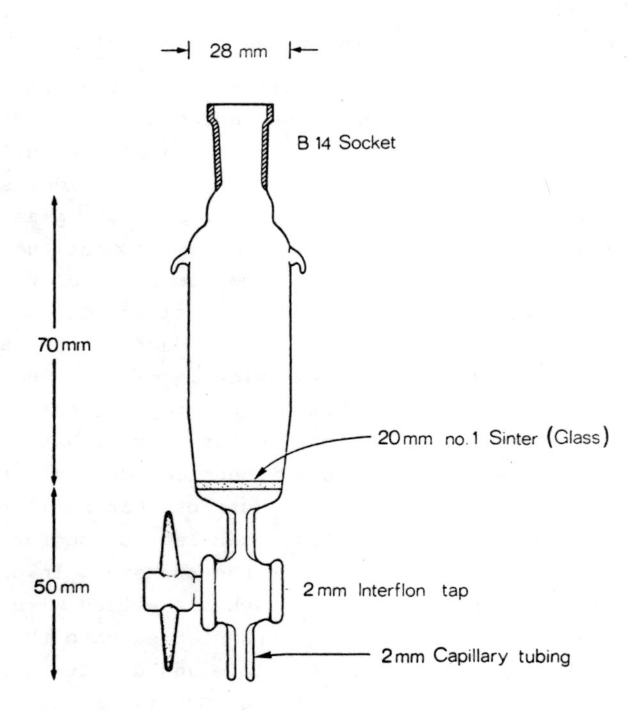

Figure 5 Resin functionalisation vessel

SIMPLIFIED PREPARATIONS OF BLOCKED 2'-DEOXYRIBONUCLEOSIDES AS STARTING MATERIALS FOR CHEMICAL OLIGONUCLEOTIDE SYNTHESIS [1]

Hans-Joachim Fritz, Wolf-Bernd Frommer, Wilfried Kramer, and Wolfgang Werr

Institut für Genetik
Universität zu Köln

SUMMARY

Gentle and convenient synthetic routes are described leading directly to 5'-monomethoxytrityl-N-acyl-2'-deoxyribonucleosides (IV) and N,3'-O-diacyl-2'-deoxyribonucleosides (XI). These compounds are key intermediates in chemical oligonucleotide synthesis and were hitherto only accessible by complicated multi-step procedures. The central feature of the new methods described in this article is transient protection of sugar hydroxy functions in 2'-deoxyribonucleosides by reaction with trimethylsilyl chloride.

1 INTRODUCTION

Chemical synthesis of oligonucleotides can be divided into three distinct phases:

A Preparation of protected monomer units (carrying up to four different blocking groups in the 2'-deoxyriboseries).

B Assembly of the oligonucleotide chain by repeated cycles of phosphorylation and terminal deprotection.

C Work-up of the final product consisting of step-wise complete deprotection, purification and characterisation.

Recent years have witnessed dramatic progress mainly with respect to part B of this scheme. Improved phosphorylation procedures [2-4] combined with immobilisation of the growing oligonucleotide chain on a solid support [4-8] have brought the time necessary to complete a full monomer addition cycle down to the order of 30 min. These developments are documented and discussed in detail in other articles within this volume.

In this situation, where the rate of the chain assembly process (which used to be by far the most tedious part of polynucleotide synthesis) is close to being limited by the speed with which the necessary reagents and solvents can be added to the reaction vessel and removed from it in a sequential manner (manually or machine-aided), further improvements of the overall process can be expected from two sides:

- Increasing the chemical selectivity of the chain elongation procedures while maintaining their speed and convenience.
- Improving the efficiency of steps A and C, i.e. preparation of protected monomers and work-up of the final product.

The present article describes simplified procedures for the preparation of double-blocked 2'-deoxyribonucleosides [step A].

A second article within this volume (H.-J. Fritz, D. Eick, and W. Werr) deals with purification and characterisation of the final product [step C].

H.G. Khorana and co-workers have elaborated a protection strategy [9] for chemical oligonucleotide synthesis which rests i.a. on the use of different, selectively removable blocking groups for the various functions of the nucleoside molecule:

- an acid-labile, substituted trityl group for the primary 5'-hydroxy function,
- an acyl group for the other terminus, i.e. the secondary 3'-hydroxy function (this blocking group can be removed by ester saponification with hydroxide ion),
- various acyl groups for the exocyclic amino functions of the nucleo-

bases A, C and G (these groups are stable to 1 M hydroxide ion, but are removed by concentrated ammonia).

This protecting group strategy, though originally developed in connection with the phosphodiester approach to polydeoxynucleotide synthesis, has proven to be so flexible that it also constitutes the basis for practically all other methods of chemical oligonucleotide synthesis presently in use.

Therefore, double protected 2'-deoxyribonucleosides of type (IV) (figure 1) (or the respective dimethoxytrityl derivatives) are important synthetic intermediates in the phosphodiester-, the phosphotriester- and the phosphite method.

A three-step scheme for the preparation of these compounds was devised many years ago by H.G. Khorana and co-workers [10,11]. This method (figure 1, route A) is still generally being used today. It proceeds by peracylating the unprotected nucleoside (I) to give intermediate (II), followed by partial de-acylation to monoacylated (III) which is tritylated to yield the final product (IV).

The procedure involves product isolation at every step and a rather harsh alkali treatment in the partial deacylation reaction.

We started to look for alternative methods for the preparation of intermediates (IV), which are chemically more gentle and which need only one-pot reactions to be carried out.

Legend to figure 1

Schematic outline of synthetic routes leading to 5'-monomethoxytrityl-N-acyl-2'-deoxyribonucleosides (IV) and N,3'-O-diacyl-2'-deoxyribonucleosides (XI), representing key intermediates in chemical oligodeoxyribonucleotide synthesis. B represents any of the three nucleobases cytosine, adenine and guanine, the RC(O)-suffix any of the three acyl groups 4-anisoyl-, benzoyl- or isobutyryl-, respectively. MMTr stands for the monomethoxytrityl group, trisubstituted silyl residues for the trimethylsilyl group. Compounds (II) are triacylated in the cases of anisoylation of 2'-deoxycytidine and isobutyrylation of 2'-deoxyguanosine; tetraacylated, in contrast, in the case of benzoylation of 2'-deoxyadenosine using benzoyl chloride [10]. In routes B and C, which do not involve partial deacylation procedures, benzoyltrifluoroacetate is used, therefore, for the direct monobenzoylation of the adenine ring [12]. For additional information refer to text.

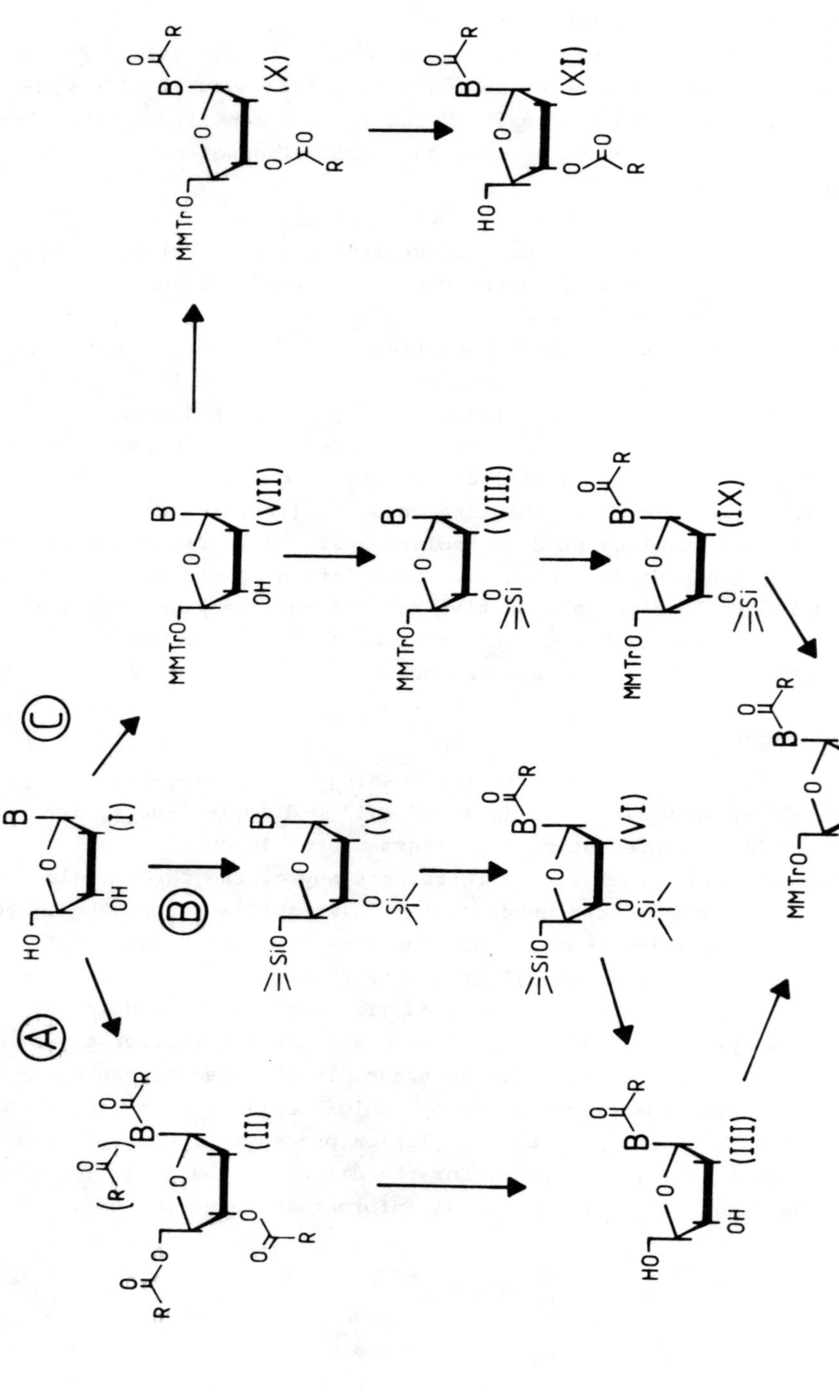
A
B
C
(I)
(II)
(III)
(IV)
(V)
(VI)
(VII)
(VIII)
(IX)
(X)
(XI)
MMTrO

2 RESULTS AND DISCUSSION

Figure 1 illustrates two synthetic schemes (B and C) which meet the requirements outlined at the end of the preceeding paragraph.

The basic idea of the approach is to make the acylation reaction N-specific by prior protection of free hydroxy functions using a silyl group.

With similar aims in mind, the <u>t-butyldiphenylsilyl</u> group has been used successfully with 2'-deoxyribonucleoside-5'-monophosphates in the phosphodiester approach by Jones, Fritz and Khorana [12]. In that case, a silyl group was required, which was stable to alkali and to the conditions of the phosphodiester condensation reaction.

Routes B and C (figure 1), in contrast, call for <u>transient</u> silylation only. Therefore, a very reactive silylation agent (leading to unstable O-Si bonds) can be applied. As a starting point, we decided to explore the use of trimethylsilyl chloride.

In the following, routes B and C will be considered individually.

<u>Route B</u> ("Silylation first"):

Unprotected 2'-deoxyribonucleoside (I) is treated with an excess of trimethylsilyl chloride in pyridine. Acylating agent is added to the resulting solution of (V) to yield (VI). (Acylating agents used were anisoyl chloride for 2'-deoxycytidine [10], isobutyric anhydride for 2'-deoxyguanosine [11] and benzoyl trifluoroacetate [13] for 2'-deoxyadenosine [12]).

Intermediate (VI) is desilylated to give (III) by addition of water. Excess water is then removed by repeated evaporations with pyridine. Finally, crude (III) (taken up again in pyridine) is tritylated to give final (IV), which is purified by silica gel column chromatography.

We found the method to work quite well for deoxycytidine and deoxyguanosine, even though with the latter the yields were somewhat variable; we were not able yet to determine the reason for this behaviour. At present, we have only preliminary evidence, that the method gives good results with 2'-deoxyadenosine.

The silyl groups of intermediates (V) and (VI) are of markedly different stability, depending on the nature of the nucleobase and its protection state: Base unprotected derivatives (V) and the fully protected derivative (VI) of 2'-deoxyguanosine are so unstable, that, in our hands, they always gave rise to several spots on tlc, seemingly indicating incomplete silylation. That this was not the case, however, was proven indirectly by the good homogeneity of derivatives (III).

<u>Route C</u> ("Tritylation first"):

Unprotected 2'-deoxyribonucleoside (I) is treated with a slight excess of 4'-monomethoxytrityl chloride in dry pyridine to give compound

(VII). The remaining hydroxy function is then blocked by addition of trimethylsilyl chloride. Resulting (VIII) is acylated to give (IX) using the same acylating agents as with route B. Addition of water to the solution of (IX) results in the formation of (IV), which is purified by silica gel column chromatography.

The method gives good results with 2'-deoxycytidine and - according to preliminary evidence - also with 2'-deoxyadenosine.

We were not able to use the method with 2'-deoxyguanosine, since our attempts of direct monomethoxytritylation of 2'-deoxyguanosine failed.

A ramification of route C leads to N,3'-O-diacylated 2'-deoxyribonucleotides (XI), which can be used as 3'-terminal residues in homogeneous phase phosphotriester syntheses: 2.5 equivalents of acylating agent are added to intermediate (VII) to give compound (X), which is smoothly detritylated to (XI) by the action of zinc bromide in anhydrous nitromethane [14,15]. So far, we have used this procedure only with 2'-deoxycytidine. The resulting N,3'-O-dianysoyl-2'-deoxycytidine was used in the synthesis of dGAAGCTTC [16].

3 PROCEDURES

3.1 5'-Monomethoxytrityl-4-anisoyl-2'-deoxycytidine by route B

2'-Deoxycytidine (2.2 mmol, 500 mg) was dried by three evaporations under reduced pressure with 20 mL portions of anhydrous pyridine. The residue was re-suspended in 25 mL of dry pyridine. Trimethylsilyl chloride (7.9 mmol, 1 mL) was added and the mixture was stirred at 20°C for 2.5 h. The product displayed three spots on tlc (silica gel, chloroform:methanol, 5:1, v/v). p-Anisoyl chloride (3.75 mmol, 0.5 mL) was added to the clear solution in two aliquots and the reaction was allowed to proceed overnight at 20°C. The product was homogeneous on tlc (see above). Water (20 mL) was added slowly to avoid precipitation. Stirring was continued for 30 min, by which time complete desilylation was confirmed by tlc (see above). The mixture was evaporated to a gum, co-evaporated several times with dry pyridine, then taken up in 25 mL of dry pyridine. Monomethoxytrityl chloride (2.5 mmol, 0.77 g) was added and the mixture was stirred overnight at 20°C. The predominant spot on tlc (see above) corresponded to the desired 5'-monomethoxytrityl-4-anisoyl-2'-deoxycytidine. The mixture was partitioned between chloroform and half-saturated aqueous sodium bicarbonate. The organic phase was evaporated to an oil and co-evaporated three times with toluene to remove excess pyridine. The resulting gum was taken up in chloroform, methanol, pyridine (100:10:0.5, v/v), then subjected to column chromatography (silica gel) using the same solvent mixture as the eluent. Fractions containing the desired material were pooled and evaporated. The product was taken up in ethyl acetate and precipitated from ether:hexane (1:1, v/v). Yield was 0.74 g, corresponding to 53%.

3.2 5'-Monomethoxytrityl-4-anisoyl-2'-deoxycytidine by route C

2'-Deoxycytidine (4.4 mmol, 1 g) was dried by several co-evaporations with anhydrous pyridine. To the suspension in dry pyridine (50 mL) was added monomethoxytrityl chloride (5 mmol, 1.54 g). The mixture showed one major trityl-positive spot (perchloric acid spray) on tlc (silica gel, chloroform:methanol, 10:1, v/v) together with some faster moving material indicating over-tritylation. The mixture was divided into two equal portions, one of which was set aside for double anisoylation (see paragraph 3.3). Trimethylsilyl chloride (6.6 mmol, 0.56 mL) was added to the other half. The reaction was allowed to proceed for 3 h at 20°C. Complete silylation was confirmed by tlc (see above). p-Anisoyl chloride (2.5 mmol, 0.33 mL) was added; the mixture was stirred overnight at 20°C. Slow addition of water (20 mL) resulted in a clear solution (some pyridine hydrochloride had precipitated before). Desilylation was complete within 30 min as judged by tlc (see above). The product was worked up by partitioning between chloroform and half-saturated aqueous sodium bicarbonate, followed by column chromatography as described in paragraph 3.1. Yield was 1.14 g, corresponding to 82%.

3.3 N,3'-O-Dianisoyl-2'-deoxycytidine

p-Anisoyl chloride (6 mmol, 0.8 mL) was added to a solution of 5'-monomethoxytrityl-2'-deoxycytidine (nominally 2.2 mmol; refer to paragraph 3.2 for details of preparation) in 25 mL dry pyridine. After 3 h at 20°C, the anisoylation reaction was complete as judged by tlc (silica gel, chloroform:methanol, 10:1, v/v). The mixture was partitioned between chloroform and half-saturated aqueous sodium bicarbonate. The organic phase was evaporated to a gum, co-evaporated several times with toluene and finally taken up in chloroform:methanol (100:1, v/v). The nucleoside was freed of residual pyridine by filtration through a bed of silica gel. The solution was evaporated, taken up in ethyl acetate and added dropwise to dry hexane (1 L) with stirring. The precipitate was collected and dried under vacuum. The fully protected nucleoside was taken up in 50 mL dry nitromethane and the solution evaporated to a gum. This was repeated twice. Finally, the residue was dissolved in 150 mL dry nitromethane. Solid zinc bromide was added with vigorous stirring until saturation was reached. Stirring was continued for 10 min. The reaction was terminated by addition of 0.2 M aqueous triethylammonium acetate (100 mL, pH 7.0). The organic phase was washed with another 100 mL portion of 0.2 M triethylammonium acetate, then with 50 mL 10% sodium chloride. The solution was co-evaporated with pyridine, then with toluene. A dry powder of the product was obtained by lyophilization from dioxane solution. The material contained tritanol and a small amount of a trityl-positive nucleoside, both of which did not interfere with use of the product in triester coupling. Detritylation of the major compound was quantitative.

3.4 5'-Monomethoxytrityl-2-N-isobutyryl-2'-deoxyguanosine by route B

2'-Deoxyguanosine (3.75 mmol, 1 g) was dried by several co-evaporations with anhydrous pyridine, then suspended in 50 mL dry pyridine. Trimethylsilyl chloride (11.3 mmol, 0.9 mL) was added with stirring. Clear solution was achieved within 5 min. The reaction was continued overnight at 20°C. The product showed three spots on tlc (silica gel, chloroform:methanol, 10:1, v/v). Isobutyric anhydride (5.6 mmol, 0.94 mL) was added and the mixture was again stirred overnight at 20°C. tlc (see above) again showed a series of spots. Addition of water (5 mL), however, resulted in smooth desilylation to a homogeneous product as judged by tlc (see above). The mixture was evaporated to a gum several times with addition of dry pyridine. The residue was taken up in 50 mL dry pyridine and monomethoxytrityl chloride (5 mmol, 1.54 g) was added. After one night's reaction at 20°C, the mixture was partitioned between chloroform and half-saturated aqueous sodium bicarbonate. The rest of the work-up was as described in paragraph 3.1. Yield was 1.05 g, corresponding to 46%.

3.5 Analysis

The products described in the preceeding paragraphs checked favourably in the following analytical tests:

a) Chromatographic comparison with authentic material (tlc and hplc on reversed phase)
b) Spectroscopy (uv and nmr)
c) Mass Spectrometry (Field Desorption).

4 CONCLUSION

We have demonstrated the practicality of transient protection of sugar hydroxy-functions by the trimethylsilyl group in developing new synthetic routes to double-blocked 2'-deoxyribonucleosides.

5'-Monomethoxytrityl-N-acyl-2'-deoxyribonucleosides are now accessible via two alternative routes requiring only one-pot reactions to be carried out. Similarly, N,3'-O-dianisoyl-2'-deoxycytidine was synthesised in two steps.

This article gives a description of the present state of the techniques as worked out in our laboratory.

We are presently trying to extend the methods and to improve the reproducibility of results obtained with purine nucleosides, which we have found to be less than satisfactory.

In all reactions, the monomethoxytrityl group was used for protection of 5'-hydroxy functions, since zinc bromide in anhydrous nitromethane removes this group sufficiently fast at least for application to oligonucleotide synthesis by homogeneous liquid-phase phosphotriester methods [14-16].

There should be no difficulties, however, to extend the procedures to the use of the dimethoxytrityl group for 5'-protection, if this should turn out to be more desirable in polymer supported synthesis.

Material synthesised by methods described in this article, was successfully used in chemical oligonucleotide synthesis [16].

ACKNOWLEDGEMENTS

We thank Dr. Gottfried Feistner and Dipl. Chem. Johannes Hegemann for measuring nmr and mass spectra. This work was supported by Deutsche Forschungsgemeinschaft through SFB 74.

REFERENCES AND NOTES

[1] This work was presented by one of us (H.-J. F.) at the 5th Symposium on the Chemistry of Nucleic Acids Components, held at Bechyne Castle, CSSR, September 1981.
At the EMBO Practical Course "Automated Chemical and Enzymic Gene Synthesis" (Darmstadt, 1982), we were informed that a similar study has, independently, been carried out by Dr. R.A. Jones and co-workers.

[2] Reese, C.B. (1978) Tetrahedron 34, 3143

[3] Letsinger, R.L. and Lunsford, W.B. (1976) J. Amer. Chem. Soc. 98, 3655

[4] a) Matteucci, M.D. and Caruthers, M.H. (1981) J. Amer. Chem. Chem. Soc. 103, 3185
b) Beaucage, S.L. and Caruthers, M.H. (1981) Tetrahedron Lett. 22, 1859

[5] Gait, M.J., Singh, N., Sheppard, R.C., Edge, M.D., Greene, A.R., Heathcliff, G.R., Atkinson, T.C., Newton, C.R., and Markham, A.F. (1980) Nucleic Acids Res. 8, 1081

[6] Miyoshi, K. and Itakura, K. (1979) Tetrahedron Lett. 38, 3635

[7] Potapov, V., Veiko, V., Koroleva, O., and Shabarova, Z. (1979) Nucleic Acids Res. 6, 2041

[8] Crea, R. and Horn, T. (1980) Nucleic Acids Res. 8, 2331

[9] Agarwal, K.L., Yamazaki, A., Cashion, P.J., and Khorana, H.G. (1972) Angew. Chemie 84, 489

[10] Schaller, H., Weimann, G., Lerch, B., and Khorana, H.G. (1963) J. Amer. Chem. Soc. 85, 3821

[11] Büchi, H. and Khorana, H.G. (1972) J. Mol. Biol. 72, 251

[12] Jones, R.A., Fritz, H.-J., and Khorana, H.G. (1978) Biochemistry 17, 1268

[13] Emmons, W.D., McCallum, K.S., and Ferris, A.F. (1953) J. Amer. Chem. Soc. 75, 6047

[14] Kohli, V., Blöcker, H., and Köster, H. (1980) Tetrahedron Lett. 21, 2683

[15] Matteucci, M.D. and Caruthers, M.H. (1980) Tetrahedron Lett. 21, 3243

[16] Werr, W. (1981) Part of "Diplom"-Thesis, Universität zu Köln

A NEW PHOSPHORYLATION PROCEDURE FOR THE INTRODUCTION OF 3',5'-INTERNUCLEOTIDE LINKAGES: SYNTHESIS OF DNA DIMERS

Jaques H. van Boom, G.A. van der Marel, C.A.A. van Boeckel, G. Wille, and C.F. Hoyng*
Department of Organic Chemistry
Gorlaeus Laboratories
Leiden
*Genentech
South San Francisco

SUMMARY

Conversion of the bifunctional phosphorylating agent 2-chlorophenyl-phosphorodichloridate 1 with two equivalents of 1-hydroxybenzotriazole affords the bis-(1-benzotriazolyl)derivative 6. The latter reagent proved to be very effective - i.e. no activating agent was required - for the introduction of a 3',5'-internucleotide linkage between properly 5'-protected and 3'-protected deoxynucleosides. According to this procedure, valuable fully protected dimers of DNA are easily accessible. Further, no side reactions on the bases thymine or guanine could be detected in this phosphorylation procedure.

1 INTRODUCTION

In the synthesis of DNA-fragments via a phosphotriester approach [1] we may discern three distinct steps:

a) the preparation of properly protected deoxynucleosides having a free hydroxyl at the 3'- or 5'-end;
b) introduction of 3',5'-internucleotide phosphotriester linkages between properly protected deoxynucleosides;
c) deblocking of all protective groups to afford DNA fragments containing solely 3',5'-phosphodiester linkages.

At the present time reliable protective groups for the protection of the hydroxyl functions of deoxynucleosides have been introduced (e.g. 4,4'-dimethoxytrityl [2], levulinoyl [3] and 2-dibromomethylbenzoyl [4]). Further, simple procedures for the removal of protective groups, and especially so for the conversion of phosphotriesters into the required phosphodiester functions [5], have been developed. However, the methodology developed so far for the introduction of 3',5'-phosphodiester linkages between nucleosides is, relatively less advanced and rather time-consuming.

Two approaches are currently in use for the introduction of 3',5'-internucleotide phosphotriester linkages. They have in common that the required phosphotriester is formed between a properly protected 3'-phosphodiester and a nucleoside having a free 5'-hydroxyl function in the presence of an activating agent (e.g. mesitylene [6] - or 2,4,6-triisopropylbenzene-sulfonyl-3-nitro-1,2,4-triazolide [7]). However, both methods differ in the way the common 3'-phosphodiester intermediate is synthesized. Thus, in one approach a monofunctional phosphorylating agent (fig. 1, **4** [8] with R^1 = 4-t-butyl-2-chlorophenyl as persistent and R^2 = 2,2,2-tribromoethyl as temporary protective group) is used for the phosphorylation of a properly 5'-protected nucleoside (Scheme 1, **7**) to afford a 3'-phosphotriester derivative (Scheme 2, **9** with R^1 = 4-t-butyl-2-chlorophenyl). Selective removal of the temporary protective group (2,2,2-tribromoethyl) of the thus obtained triester affords the required 3'-phosphodiester intermediate. In the other approach, a bifunctional phosphorylating agent (fig. 1, **1** with R^1 = 2-chlorophenyl) is converted into 2-chlorophenyl-phosphoro di-(1,2,4--triazolide) [9] (fig. 1, **3** with R^1 = 2-chlorophenyl). An excess of the latter reagent reacts with a properly protected nucleoside (Scheme 1, **7**) to afford a monotriazolide derivative, which is converted into the required 3'-phosphodiester intermediate by hydrolysis with triethylamine:water [10]. In this paper we wish to report in detail that fully-protected dimers **14** and **15** (scheme 3) are easily accessible by

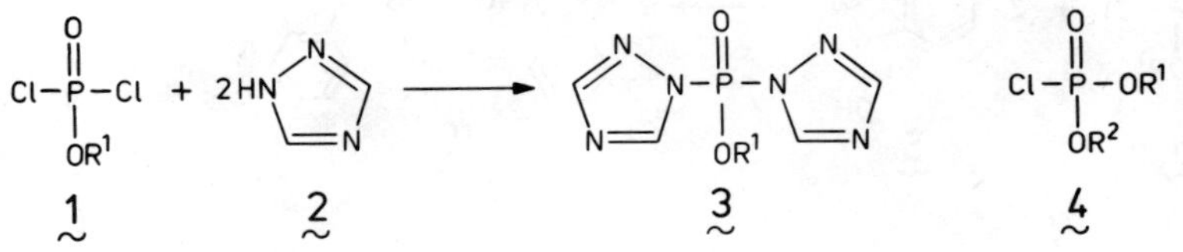

Figure 1 Conversion of the bifunctional phosphorylating agent 2-chlorophenylphosphorodichloridate with two equivalents of 1-hydroxybenzotriazole to yield the bis-(1-benzotriazolyl)derivate

phosphorylation of 5'-protected nucleosides (scheme 1, 7) with a phosphorylating agent (scheme 1, 6), which has been activated with 1-hydroxybenzotriazole. The intermediate 3'-phosphotriester derivative (scheme 1, 8) thus obtained reacts effectively with nucleosides having a free 5'-hydroxyl function (scheme 3, 13) to afford fully-protected dimers (scheme 3, 14).

2 RESULTS AND DISCUSSION

The application of 1-hydroxybenzotriazole (HOBT) to form active esters of amino acids which function as intermediates for the formation of peptide bonds is well studied [11].

In previous reports [12,13] we showed that HOBT is also an effective reagent to convert a bifunctional phosphorylating agent 1 [14] (R^1 = 2-chlorophenyl) into the active triester intermediate 6 (R^1 = 2-chlorophenyl, scheme 1). The O,O-bis-[1-benzotriazole] derivative 6 is easily accessible by reacting together phosphorylating agent 1, two equivalents of HOBT 5 and two equivalents of pyridine in peroxide-free tetrahydrofuran (THF). Removal, after 1.5 h, of the pyridine hydrochloric acid salt gives a solution of agent 6 in THF, which can be kept for several months at 0°C. The triester intermediate 6 (scheme 1) proved to be not only an effective phosphorylating agent, but it also lacked the disadvantages adherent to the use of the di-triazolide derivative 3 (fig. 1) or the monofunctional reagent 4 (fig. 1). Thus, phosphorylation of the 5'-protected nucleoside 7 (R^2 = TBDMS, scheme 1)

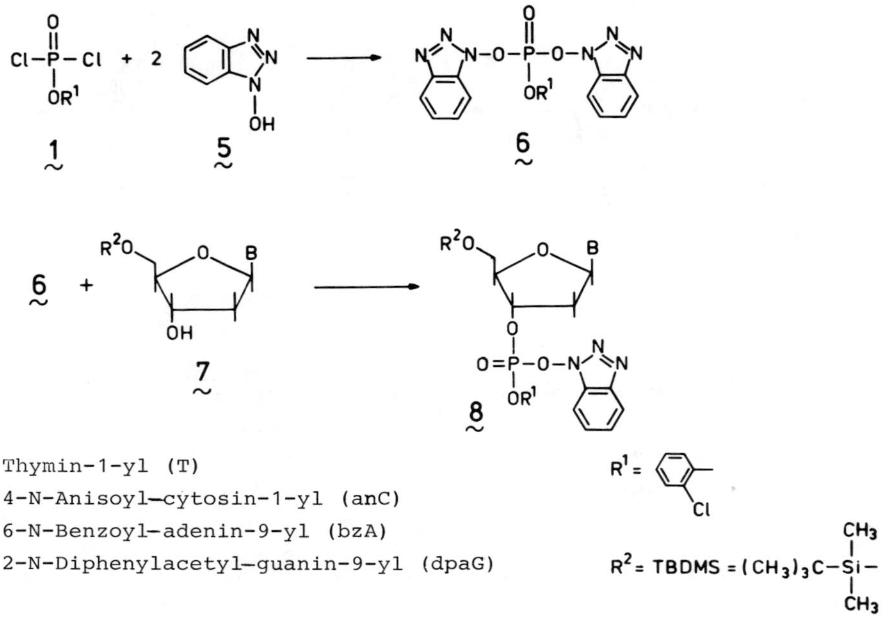

B = Thymin-1-yl (T)
B = 4-N-Anisoyl-cytosin-1-yl (anC)
B = 6-N-Benzoyl-adenin-9-yl (bzA)
B = 2-N-Diphenylacetyl-guanin-9-yl (dpaG)

R^1 = 2-chlorophenyl
R^2 = TBDMS = $(CH_3)_3C-Si(CH_3)_2-$

Scheme 1

with a slight excess of 6 and analysis, after 45 min at 20°C, of the reaction mixture showed the presence of solely baseline material on tlc. Work-up of the reaction mixture and hydrolysis of the product with aqueous triethylammonium bicarbonate (TEAB), followed by extraction with chloroform afforded as corroborated by ^1H- and ^{31}P-nmr spectroscopy the 3'-(2-chlorophenyl)-phosphate of 8 (B=T). The favourable phosphorylating properties of intermediate 8 were demonstrated in the synthesis of four 3'-phosphotriester derivatives (scheme 2, 9a, 10, 11 and 12), and also in the synthesis of two types of dimers (scheme 3, terminal dimers 14a and non-terminal dimers 15a). The nucleosides which we used in this study to prepare the 3'-phosphotriester derivatives in scheme 2, as well as the different dimers in scheme 3 were protected at the 5'-end with the t-butyldimethylsilyl group (TBDMS) [15]. The reason for this choice was as follows. Firstly, the TBDMS derivatives of all 4 deoxynucleosides 7 (B = T, bzA, anC, dpaG) could easily be obtained as stable compounds. Secondly, the efficiency of the removal of the TBDMS group was acceptable. For instance, the terminal nucleoside 13 (R^3 = benzoyl) was obtained in high yields by benzoylation of 7 (R^2 = TBDMS), followed by removal of the TBDMS group with dry tetra-n-butylammonium fluoride (TBAF) in THF [16]. On the other hand, deblocking of the TBDMS group from the dimers 14a and 15a (scheme 3) with p-toluenesulfonic acid (pTsOH) afforded, despite the occurrence of depurination, acceptable yields of deblocked dimers 14b and 15b (R^2 = H) respectively (see yields obtained of dimers 14b and 15b in table 1). Finally, a less conspicuous reason for the use of the relatively acid stable TBDMS is given by the fact that a very acid labile protective group [4,4'-dimethoxytrityl group (DMTr)] at the 5'-end of nucleosides is not stable under the conditions applied to convert 7 (R^2 = TBDMS) into intermediate 8 (R^2 = TBDMS, scheme 1). Fortunately, however, we could convert 7 (R^2 = DMTr) into derivative 8 (R^2 = DMTr) by performing the phosphorylation in the presence of pyridine (see later).

Having established the identity of intermediate 8 (B = T) and no formation of a symmetrical product (the product which may be obtained when 6 reacts twice with 7), we now turned our attention to the synthesis of 3'-phosphotriesters of nucleosides (scheme 2, 9-12), which are valuable building blocks for the synthesis of non-terminal dimers 15 in scheme 3. Thus, addition of 2,2,2-tribromoethanol to a solution of 8 (B = T, bzA, anC or dpaG) in THF gave, after work-up and purification by short-column chromatography, triester 9a (B = T, bzA, anC or dpaG) in a yield of 90%. Removal of the TBDMS group from 9a with pTsOH afforded after work-up and purification by short-column chromatography, 9b (R^2 = H, B = T, bzA, anC or dpaG) in acceptable yields. In this way, we prepared the non-terminal units 9a of all four deoxynucleosides having a 3'-phosphotriester, the 2,2,2-tribromoethyl group which can selectively be removed in the presence of the other protective group (R^1 = 2-chlorophenyl).

In the same way, we prepared the 3'-phosphotriester derivatives 11 and 12 (B = T) in good yields, by treating 8 (B = T, R^1 = 2-chlorophenyl, R^2 = TBDMS) with 4-nitro-phenylethanol or 5-chloro-8-hydroxyquinoline, respectively. We also synthesized the 3'-phosphotriester 10 (B = T, R^1 = 2-chlorophenyl, R^2 = TBDMS) by treating 8 (B = T, R^1 = 2-chlorophenyl, R^2 = TBDMS) with 3-hydroxypropionitrile, using pyridine instead of 1-methylimidazole as the tertiary base. In this respect, it is interesting to note that the use of pyridine decreases the rate of phosphorylation.

The above described hydroxybenzotriazole (HOBT) approach to the synthesis of the 3'-phosphotriester intermediates 9-12, which have proven to be valuable intermediates for the synthesis of nucleic acids, does not give rise to side-reactions and is also more efficient and economic than the methods commonly in use for the preparation of these

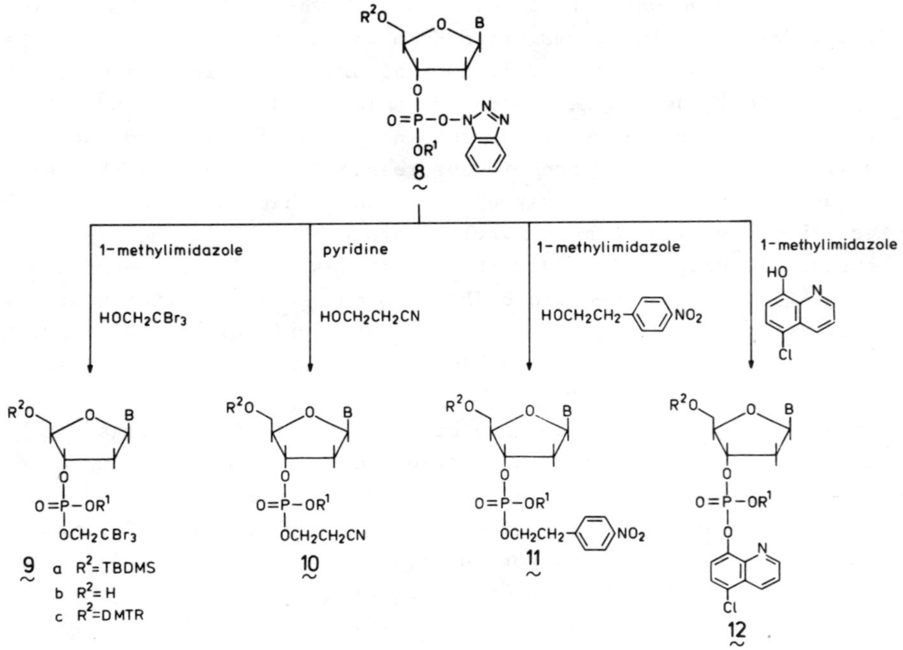

Scheme 2

compounds. For instance, the introduction of the 3'-phosphodiester function of 9 (R^1 = 4-t-butyl-2-chlorophenyl) was previously achieved by phosphorylation of a properly protected nucleoside (scheme 1, 7) with the monofunctional reagent 4 [8] (R^1 = 4-t-butyl-2-chlorophenyl, R^2 = 2,2,2-tribromoethyl in fig. 1). The two types of 3'-phosphotriester functions present in compounds 10 and 12 in scheme 2 were also prepared by phosphorylation of properly protected nucleosides with the, in situ prepared, corresponding monofunctional phosphorylating agents 4 (fig. 1). Thus, Crea et al. [17] obtaines 10 (R^1 = 4-chlorophenyl) by using agent 4 (R^1 = 4-chlorophenyl, R^2 = CH_2CH_2CN in fig. 1). On the other hand, Takaku et al. [18] obtained the 3'-phosphotriester 12 by treating properly protected ribonucleosides with agent 4 (R^1 = 4-chlorophenyl, R^2 = 5-chloro-8-quinolyl). The above described approach, using monofunctional agents 4, may lead to unwanted side-reactions on the nucleoside-base cytosine [7] and also to modification of the guanine base [19]. Furthermore, in order to effort complete phosphorylation of the starting product an excess of monofunctional phosphorylating agents, the preparation of which is rather time-consuming, had to be used. Other existing methods to the preparation of the different types of 3'-phosphotriester derivatives of nucleosides in scheme 2 are either based on the use of the ditriazolide reagent 3 (fig. 1), or on a process in which an activating agent is used. Thus, Balgobin et al. [20] introduced the 2,2,2-tribromoethyl group by treatment of a 3'-(2-chlorophenyl)phosphate of a properly protected dimer with 2,2,2-tribromoethanol in the presence of the activating agent 1-mesitylenesulfonyl-3-nitro-1,2,4-triazolide. Takaku et al. [21] also succeeded in preparing the 3'-phosphotriester function as present in compound 12 (R^1 = CH_2CH_2CN) by activating a properly protected ribonucleoside 3'-(5-chloro-8-quinolyl)phosphate with 8-quinolylsulfonyl chloride in the presence of 3-hydroxypropionitrile.

It is quite clear that a methodology which is based on the use of an activating agent is, in comparison with the HOBT approach, rather laborious. On the other hand, Broky et al. [22] succeeded in preparing intermediate 10 (R^1 = 4-chlorophenyl, R^2 = DMTr) by reacting 7 (R^2 = DMTr) with the ditriazolide reagent 3, originally introduced by Narang et al. [9] followed by the addition of 3-hydroxypropionitrile in the presence of 1-methylimidazole. Using the same ditriazolide procedure, Pfleiderer et al. [23] prepared the 3'-phosphotriester function 11 (R^2 = 2-chlorophenyl) using 4-nitrophenylethanol instead of 3-hydroxypropionitrile. Although the ditriazolide approach presents a shorter route to the 3'-phosphotriester derivatives 9-12 (scheme 2) than the

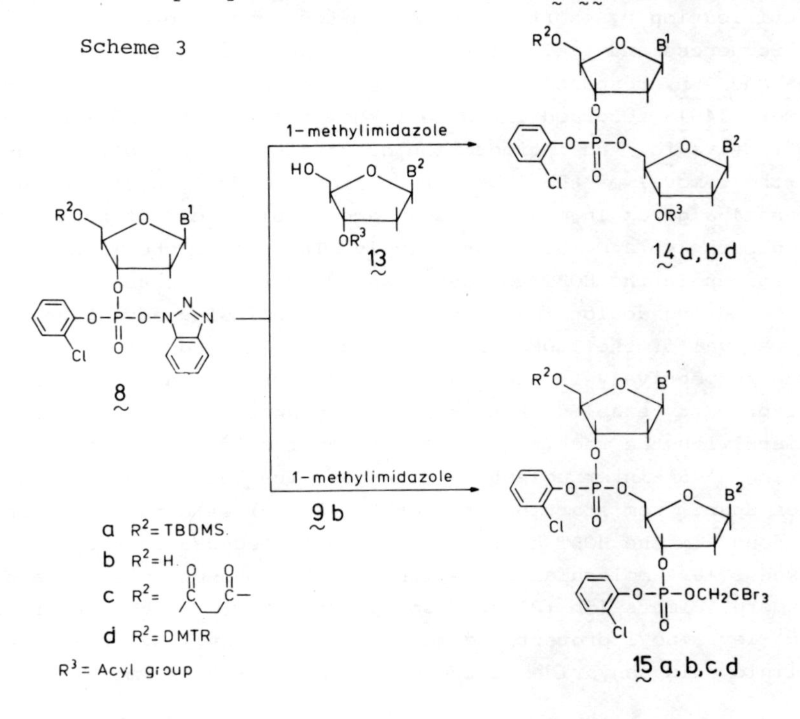

one based on the use of activating agents, it is, for the reasons given below, less superior than the HOBT approach. Thus, in order to prevent the formation of symmetrical products an excess (1.4 [22] to 2.4 [4] molecular equivalents) of the ditriazolide 3 has to be used. Further, modification of the nucleoside bases thymine [24], and especially so guanine [25] is a serious disadvantage to this method.

The HOBT approach proved also to be very effective for the introduction of 3',5'-internucleotide linkages (scheme 3). Thus addition of a 3'-protected nucleoside 13 (R^3 = benzoyl; 0.85 mmol) to a solution of HOBT intermediate 8 in THF, which is obtained by the reaction of 6 (1.1 mmol) with 7 (1.0 mmol), gave in the presence of 1-methylimidazole (2.5 mmol), after 1.5 h at 20°C and work-up of the reaction mixture followed by purification, the fully-protected terminal dimer 14a (R^2 = TBDMS, R^3 = Bz). In this way, we prepared four fully-protected dimers in a yield of 70 to 90% (table 1). Dimers 14b, having a free 5'-OH group were isolated in acceptable yields (table 1) after performing purification by short column chromatography of crude dimers 14b (R^2 = H), which were obtained by deblocking of the TBDMS group from the corresponding crude dimers 14a (R^2 = TBDMS) with pTsOH. According to the same procedure we also prepared the non-terminal dimers 15a (R^2 = TBDMS) by phosphorylation of 8 with the non-terminal unit 9b (scheme 3). Removal of the TBDMS group from crude dimers 15a afforded, after work-up and purification by short column chromatography pure dimers 15b (R^2 = H). Sequences and yields of the fully- and partially protected dimers 15a and 15b, respectively, are recorded in table 1. The yields of the dimers 14-15 recorded in table 1 show that: a) the HOBT approach is very effective for the introduction of 3',5'-internucleotide linkages; b) the removal of the TBDMS group from the fully protected dimers 14a and 15a gives in most cases an acceptable yield of partially protected dimers 14b and 15b, respectively. The use of the TBDMS as a protective group in the HOBT approach has, despite the occurrence of depurination during deblocking with acid, the following advantages. Thus, the removal of the TBDMS group from crude 14a and 15a to give 14b and 15b respectively is accompanied by a decrease in R_f-values. The latter property enabled us to separate by short column chromatography dimers with the required 3',5'-internucleotide linkages from dimers having 5',5'-internucleotide linkages. The latter unwanted dimers, which apart from starting product 8 (R^2 = H) were the sole impurities we found in the HOBT approach, may be formed due to the use of a small excess (0.1 molecular equivalent) of the phosphorylating agent 6. Furthermore, dimers 15b (R^2 = H) could quantitatively be converted into the 5'-levulinoyl protected dimers 15c (R^2 = lev.) by treatment with levulinic acid anhydride in the presence of 1-methylimidazole.

The latter dimers proved to be very convenient starting products for the synthesis of DNA fragments with a defined sequence and length.

Finally, we also showed that the non-terminal dimers 15d, which are protected at the 5'-end with the very acid-labile 4,4'-dimethoxytrityl group were equally well accessible by a slight modification of the first phosphorylation step of the HOBT approach (phosphorylation of 7 (R^2 = DMTr) with 6 to give 8 (R^2 = DMTr) in scheme 1). Thus, addition of pyridine or 2,6-lutidine to the reaction mixture of the first phosphorylation step afforded, after applying the same conditions and purification procedure as given for the synthesis of dimers 15a (R^2 = TBDMS) in scheme 3, the fully protected dimers 15d (R^2 = DMTr), the sequences and yields of which are recorded in table 1. Also in this case, we were able to remove effectively small impurities of unwanted dimers containing 5',5'-phosphotriester linkages by short-column chromatography. The homogeneity and identity of the dimers recorded in the table were corroborated by deblocking of the fully-protected products by the following procedures. Thus, dimers 14a and 15a were firstly converted into 14b and 15b respectively. After levulinoylation, the two types of dimers 14c and 15c thus obtained, were subjected to the following deblocking and analysis procedures. Thus dimers 14c (R^2 = lev) were first treated with N^1,N^1,N^3,N^3-tetramethylguanidinium syn--pyridine-2-carboxaldoximate (oximate) [26], followed with aqueous ammonia for 20 h at 50°C. The crude dimers were purified by anion-exchange chromatography on a column of DEAE Sephadex A25 and digested with spleen phosphodiesterase. In all cases, hplc analysis of the digest showed complete digestion of the dimers and solely the presence of the deoxynucleosides and deoxynucleosides-3'-phosphates. The non--terminal dimers 15c (R^2 = lev) were first treated with oximate, followed by aqueous ammonia and finally, purified by DEAE Sephadex A25 anion-exchange chromatography. hplc analysis of the digest showed complete digestion of the dimers and the presence of solely the expected deoxynucleosides-3'-phosphates and deoxynucleosides-3'-(2,2,2-tribromoethyl)phosphates. Dimers 15d were deblocked according to the same procedure followed for the deblocking of dimers 15c (R^2 = lev) except that in the last step the DMTr group was removed by short treatment with aqueous acetic acid. Also in these cases hplc analysis of the spleen phosphodiesterase digests showed complete digestion of the dimers into solely deoxynucleoside-3'-phosphates and deoxynucleoside--3'-(2,2,2-tribromoethyl)phosphates. In conclusion the data presented in this paper clearly show that the HOBT approach gives an easy access to well defined dimers. The only side-reaction adherent to the use of this approach is the formation of small quantities of dimers containing 5',5'-internucleotide linkages. Thus, side-reactions which may

Table 1 Yields of dimers 14 and 15 synthesized[a] according to scheme 3

dimer 14a No	R^2	B^1	B^2	yield[b]	dimer 14b No	R^2	B^1	B^2	yield[d]
1	TBDMS	anC	dpaG	80	1	H	T	anC	75
2	"	T	T	93	2	H	T	dpaG	90
3	"	bzA	T	76	3	H	dpaG	dpaG	76
4	"	dpaG	T	72	4	H	anC	anC	72

dimer 15a No	R^2	B^1	B^2	yield[c]	dimer 15b No	R^2	B^1	B^2	yield[e]
1	TBDMS	T	T	88	1	H	T	dpaG	80
2	"	T	anC	78	2	H	T	bzA	61
3	"	anC	T	77	3	H	anC	anC	85
4	"	anC	dpaG	88	4	H	anC	bzA	59
5	"	dpaG	anC	75	5	H	dpaG	T	78
6	"	dpaG	bzA	90	6	H	dpaG	dpaG	77
7	"	bzA	T	80	7	H	bzA	anC	69
8	"	bzA	dpaG	85	8	H	bzA	bzA	50

dimer 14d No	R^2	B^1	B^2	yield[c]	dimer 15d No	R^2	B^1	B^2	yield[c]
1	DMTr	m^5bzC	dpaG	81	1	DMTr	T	bzA	63
2	"	m^5bzC	m^5bzC	88	2	"	T	dpaG	75
3	"	anC	m^5bzC	84	3	"	T	T	80
4	"	dpaG	bzA	69	4	"	T	anC	74

a For details see experimental part

b After short-column chromatography, based on 5'-hydroxy component 13

c After short-column chromatography, based on 5'-hydroxy component 9b

d After removal of the TBDMS (R^2) group from crude dimer 14a and short-column chromatography, based on 5'-hydroxy component 13

e After removal of the TBDMS (R^2) group from crude dimer 15a and short-column chromatography, based on 5'-hydroxy component 9b lead to the introduction of modification of the nucleoside bases guanine, thymine and also uracil [12] were not observed. We therefore believe that this approach is superior than the recently introduced modification of the ditriazolide approach to the synthesis of dimers by Broka et al. [22] and also Agarwal et al. [27]; in both approaches the formation of side-products leading to extensive modification of the base guanine is not excluded.

3 EXPERIMENTAL

3.1 General methods and materials

Pyridine and tetrahydrofuran were dried by refluxing with CaH_2 for 16 h and then distilled. Pyridine was redistilled from p-toluenesulfonyl chloride (60 g/L) and stored over molecular sieves 4 Å. Tetrahydrofuran was redistilled from $LiAlH_4$ (5 g/L) and stored over molecular sieves 5 Å.

1-Methylimidazole and 3-hydroxypropionitrile were distilled under reduced pressure and stored over molecular sieves 4 Å.

4-Nitrophenylethanol, 5-chloro-8-hydroxyquinoline, 2,2,2-tribromoethanol, and 1-hydroxybenzotriazole were purchased from Aldrich.

1-Hydroxybenzotriazole was dried _in vacuo_ (P_2O_5) for 70 h at 50°C.

Schleicher and Schüll tlc sheets F 1500 LS 254 were used for tlc in solvent systems A (chloroform:methanol; 92:8, v/v) and B (chloroform: methanol; 85:15, v/v). Short column chromatography was performed on kieselgel 60 (230-400 mesh ASTM) suspended in chloroform.

^{1}H-nmr spectra of the protected nucleotides were measured at 100 MHz with a Jeol JNMPS 100 spectrometer; shifts are given in ppm (δ) relative to tetramethylsilane (TMS) as internal standard.

^{31}P-nmr spectra of the protected nucleotides were measured at 40.48 MHz with a Jeol JNMPFT 100 spectrometer equipped with an EC-100 computer, operating in the Fourier transform mode; shifts are given in ppm (δ) relative to 85% H_3PO_4 as external standard.

3.2 Synthetic procedures

3.2.1 Synthesis of 2-chlorophenyl-bis-[O,O-benzotriazolyl]phosphate 6 (R^1 = 2-ClC_6H_4)

A solution of 2-chlorophenylphosphorodichloridate 1 (R^1 = 2-ClC_6H_4; 5.0 mmol, 1.22 g) in 5 mL anhydrous tetrahydrofuran was added dropwise to a stirred and cooled (ice-water bath) solution of 1-hydroxybenzotriazole 5 (10.0 mmol, 1.35 g) and pyridine (10.0 mmol, 0.8 mL) in

20 mL anhydrous tetrahydrofuran. The ice-water bath was removed and, after another 1.5 h, the precipitate was filtered off under anhydrous conditions. The thus obtained solution of phosphorylating agent 6 was used immediately.

3.2.2 Synthesis of 5'-O-butyldimethylsilyldeoxythymidine 3'-O-(2-chlorophenyl)phosphate

A solution of phosphorylating agent 6 (R^1 = 2-ClC_6H_4; 1.1 mmol, 486 mg) in 5.5 mL tetrahydrofuran was added, under anhydrous conditions, to 5'-O-t-butyldimethylsilyldeoxythymidine 7 (B = T, R^2 = TBDMS; 1.0 mmol). The mixture was stirred for 25 min at 20°C. tlc analysis (system B) of the reaction mixture showed that 7 (B = T, R^2 = TBDMS) was completely converted into base-line material. The reaction mixture was diluted with 50 mL chloroform, washed with triethylammoniumbicarbonate (TEAB, 3 x 25 mL). The organic layer was concentrated and triturated with 200 mL petroleum ether (40-60°C). The residue was dissolved in chloroform and concentrated to a glass. ^{31}P-nmr ($CDCl_3$); -7.13. ^{1}H-nmr ($CDCl_3$); 6.26 (H_1', t), 4.84 (H_3', m), 4.18 (H_4', m), 1.80 (CH_3, s).

3.2.3 General procedure for the phosphorylation of partially protected nucleoside 7

Step 1:

Preparation of intermediate 8 (R^1 = 2-ClC_6H_4, R^2 = TBDMS, B = T, bzA, anC or dpaG). A solution of phosphorylating agent 6 (R^1 = 2-ClC_6H_4; 1.2 mmol, 531 mg) in 6.0 mL tetrahydrofuran was added, under anhydrous conditions, to nucleoside 7 (R^2 = TBDMS, B = T, bzA, anC or dpaG; 1.0 mmol). The mixture was stirred for 25 min at 20°C. tlc analysis (system B) of the reaction mixture showed that nucleoside 7 (R^2 = TBDMS, B = T, bzA, anC or dpaG) was completely converted into base-line material. The thus obtained solution of intermediate 8 was immediately used for the synthesis of compounds 9,10,11 and 12 (scheme 2).

Step 2:

Preparation of deoxynucleotide 9 (R^1 = 2-ClC_6H_4, R^2 = TBDMS, B = T, bzA, anC or dpaG). To a solution of intermediate 8 (R^2 = TBDMS, B = T, bzA, anC or dpaG; 1.0 mmol) in tetrahydrofuran (6.0 mmol) was added 2,2,2-tribromoethanol (1.5 mmol, 424 mg) and, after 10 min, 1-methylimidazole (2.5 mmol, 0.2 mL). tlc analysis (system A), after 1 h at 20°C, showed the reaction to be complete. The reaction mixture was diluted with 75 mL chloroform, washed twice with 30 mL 1.0 M TEAB and water (30 mL). The organic layer was dried ($MgSO_4$) and concentrated to an oil. The latter was dissolved in 2.0 mL chloroform and applied to a column of kieselgel (14 g) suspended in chloroform. Elution of the column was effected with chloroform:methanol (98:2, v/v). The appropriate fractions were concentrated to give a glass.

5'-O-t-butyldimethylsilyldeoxythymidine 3'-O-(2,2,2-tribromoethyl 2-chlorophenyl) phosphate

Compound 9a (R^1 = 2-ClC_6H_4, R^2 = TBDMS, B = T) was obtained as a mixture of diastereoisomers in a yield of 90%. ^{1}H-nmr ($CDCl_3$); 6.36 (H_1', dd), 5.30 (H_3', m) 5.08 (CH_2CBr_3), 0.9 (t-butyl, s).

5'-O-t-butyldimethylsilyl-4-N-anisoyldeoxycytidine 3'-O-(2,2,2-tribromoethyl 2-chlorophenyl)phosphate

Compound 9a (R^1 = 2-ClC_6H_4, R^2 = TBDMS, B = anC) was obtained as a mixture of diastereoisomers in a yield of 90%. ^{1}H-nmr ($CDCl_3$); 8.27 (H_6, d, 8 Hz), 7.93 and 6.95 (anisoyl, d, 9 Hz), 6.36 (H_1', t), 4.83 (CH_2CBr_3), 0.90 (t-butyl, s). ^{31}P-nmr ($CDCl_3$); -10.18 and -10.34.

5'-O-t-butyldimethylsilyl-6-N-benzoyladenosine 3'-O-(2,2,2-tribromoethyl 2-chlorophenyl)phosphate

Compound 9a (R^1 = 2-ClC_6H_4, R^2 = TBDMS, B = bzA) was obtained as a mixture of diastereoisomers in a yield of 90%. ^{1}H-nmr ($CDCl_3$); 8.80 (H_8, s), 6.32 (H_1', t), 5.56 (H_3', m), 4.92 (CH_2CBr_3), 0.90 (t-butyl, s). ^{31}P-nmr ($CDCl_3$); -10.07 and -10.20.

5'-O-t-butyldimethylsilyl-2-N-diphenylacetyldeoxyguanosine 3'-O-(2,2,2--tribromoethyl 2-chlorophenyl)phosphate

Compound 9a (R^1 = 2-ClC_6H_4, R^2 = TBDMS, B = dpaG) was obtained as a mixture of diastereoisomers in a yield of 90%. ^{1}H-nmr ($CDCl_3$); 7.92 (H_8, s), 6.22 (H_1', t), 5.50 (H_3', m), 5.38 (CH, dpa, s), 4.81 (CH_2CBr_3). ^{31}P-nmr ($CDCl_3$); -9.90.

5'-O-t-butyldimethylsilyldeoxythymidine 3'-O-(2-cyanoethyl 2-chlorophenyl)phosphate 10 (B = T)

The same procedure described for the synthesis of product 9a was applied for the preparation of 10 (R_1 = 2-ClC_6H_5, R^2 = TBDMS, B = T). Thus, 3-hydroxypropionitrile (3.0 mmol, 213 mg) and pyridine (3.0 mmol, 0.24 mL) were used instead of 2,2,2-tribromoethyl and 1-methylimidazole. Reaction time: 2.5 h. Deoxynucleotide 10 (R^1 = 2-ClC_6H_5, R^2 = TBDMS, B = T) was obtained as a mixture of diastereoisomers in a yield of 92%.

^{1}H-nmr ($CDCl_3$); 7.53 (H_6, s), 6.41 (H_1', t), 5.24 (H_3', m), 4.45 (OCH_2), 3.04 (CH_2CN, t). ^{31}P-nmr ($CDCl_3$); -8.86.

5'-O-t-butyldimethylsilyldeoxythymidine 3'-O-(4-nitrophenylethyl 2-chlorophenyl)phosphate 11 (B = T)

The same procedure described for the synthesis of product 9a was applied for the preparation of 11 (R^1 = 2-ClC_6H_4, R^2 = TBDMS, B = T). Thus, 4-nitro-phenylethylalcohol (2.0 mmol, 334 mg) was used instead of 2,2,2-tribromoethanol. Reaction time: 1.5 h. Compound 11 (R^1 = 2-ClC_6H_4, R^2 = TBDMS, B = T) was obtained as a mixture of diastereoisomers in a yield of 92%. ^{1}H-nmr ($CDCl_3$); 6.43 (H_1', t) 5.21 (H_3', m), 4.57 (OCH_2), 3.17 ($CH_2C_6H_4NO_2$, t). ^{31}P-nmr ($CDCl_3$); -8.23 and -8.34.

5'-O-t-butyldimethylsilyldeoxythymidine 3'-O-(5-chloroquinoline 2-chlorophenyl)phosphate 12 (B = T)

The same procedure described for the synthesis of product 9a was applied for the preparation of 12 (R^1 = 2-ClC_6H_4, R^2 = TBDMS, B = T). Thus 5-chloro-8-hydroxyquinoline (2.5 mmol, 450 mg) was used instead of 2,2,2-tribromoethanol. Reaction time: 1.5 h. Compound 12 (R^1 = 2-ClC_6H_4, R^2 = TBDMS, B = T) was obtained as a mixture of diastereoisomers in a yield of 64%. ^{1}H-nmr ($CDCl_3$); 6.53 (H_1', t) 5.63 (H_3', m), 1.94 (CH_3, s). ^{31}P-nmr ($CDCl_3$); -13.39 and -13.63.

3.2.4 General procedure for the preparation of fully-protected dimers 14 (R^2 = TBDMS)

A solution of phosphorylating agent 6 (R^1 = 2-ClC_6H_4; 1.1 mmol, 486 mg) in 5.5 mL tetrahydrofuran was added, under anhydrous conditions, to nucleoside 7 (R^1 = TBDMS, B = T, bzA, anC or dpaG; 1.0 mmol). The mixture was stirred for 25 min at 20°C. tlc analysis (system B) of the reaction mixture showed that nucleoside 7 (R^2 = TBDMS, B = T, bzA, anC or dpaG) was completely converted into base-line material. Deoxynucleoside 13 (B = T, bzA, anC or dpaG; 0.85 mmol) was now added and after 10 min, 1-methylimidazole (2.5 mmol, 0.2 mL). tlc analysis (system A and B) of the reaction mixture, after 1.5 h, showed the reaction to be complete. The reaction mixture was diluted with 75 mL chloroform and washed twice with 30 mL 1 M TEAB and water (30 mL). The organic layer was dried ($MgSO_4$) and concentrated to a glass. The thus obtained crude dimer 14a (R^2 = TBDMS) was converted into dimer 14b (R^2 = H) by the selective removal of the TBDMS protective group (see next section) or was purified by short-column chromatography. Thus, a solution of dimer 14a (R^2 = TBDMS) in 2.0 mL chloroform was applied to a column of kieselgel (14 g) suspended in chloroform. Elution of the column was effected with chloroform:methanol (98-96:2-4, v/v). The appropriate fractions were collected and concentrated to give a glass. Data on the yields of pure dimers 14a and 14b are summarised in table 1.

3.2.5 General procedure for the preparation of dimer 15 (R^2 = TBDMS)

The same procedure described for the synthesis of dimer 14a was applied for the preparation of dimer 15a (R^2 = TBDMS). This nucleotide 9b (0.85 mmol) was used instead of nucleoside 13. Data on the yields of pure dimers 15a and 15b are summarised in table 1.

3.2.6 General procedure for the preparation of dimers 14d and 15d (R^2 = DMTr)

The preparation of dimer 14d (R^2 = DMTr) and 15d (R^2 = DMTr) was performed according to a slight modification of the procedure described for the synthesis of dimer 14a (R^2 = TBDMS). Thus, the first step of the phosphorylation of 7 (R^2 = DMTr; 1.0 mmol) in 5.5 mL tetrahydrofuran with reagent 6 (1.1 mmol) was performed in the presence of pyridine (1.1 mmol, 0.09 mL) or 2,6-lutidine (1.1 mmol, 128 μL). Data on the yield of pure dimers 14d and 15d are summarised in table 1.

3.2.7 General procedure for the levulionylation of partially protected dimers 14b and 15b

To a solution of dimer 14b or 15b (1.0 mmol) in 1.0 mL anhydrous pyridine was added a solution of levulinic acid anhydride (1.5 mmol, 321 mg) in 3.0 mL anhydrous dioxane. The reaction mixture was cooled (ice-water bath) and 1-methylimidazole (0.5 mmol, 0.04 mL) was added. After 1.5 h tlc analysis (system A and B) of the reaction mixture showed the reaction to be complete. The reaction was stopped with 0.1 mL water. The mixture was concentrated to an oil. The latter was dissolved in 75 mL chloroform, washed with 10% $NaHCO_3$ (30 mL) and water (30 mL). The organic layer was dried ($MgSO_4$) and concentrated to an oil. The crude dimers (i.e. 14c or 15c) were dissolved in 2 mL chloroform and applied to a column of kieselgel (15 g). The column was eluted with chloroform-methanol (97.5-96:2.5-4, v/v). The appropriate fractions were concentrated and precipitated from petroleum ether (40-60°C). The precipitate was filtered off and stored *in vacuo* (P_2O_5). Pure dimers 14c and 15c were obtained in a yield of 90%.

3.2.8 Removal of protecting groups

3.2.8.1 General procedure for the removal of the TBDMS-group from fully-protected mono- and dinucleotides

To a stirred solution of a fully-protected monomer (9a, 10a, 1.0 mmol) or dimer (14a, 15a, 1.0 mmol) in 8.0 mL acetonitrile and 2.0 mL water was added p-toluenesulfonic acid (5.0 mmol, 0.81 g). tlc analysis (system A and B), after 0.5-1.0 h, showed the reaction to be complete. The reaction mixture was diluted with 50 mL chloroform and 10% $NaHCO_3$ (30 mL) was added. The organic layer was washed with 30 mL water, dried ($MgSO_4$) and concentrated to an oil. The latter was triturated with 200 mL petroleum ether (40-60°C). The residue was dissolved in 2.0 mL chloroform and applied to a column of kieselgel (14 g) suspended in chloroform. Elution of the column was effected with chloroform:methanol (98-96:2-4, v/v). The appropriate fractions were concentrated and precipitated from 200 mL petroleum ether (40-60°C). The precipitate was filtered off and dried *in vacuo* (P_2O_5). Data on the yields of pure dimers 14b and 15b are summarised in table 1.

3.2.8.2 Deblocking of fully-protected dimers

A solution of N^1,N^1,N^3,N^3-tetramethylguanidinium syn-pyridine-2--carboxaldoximate (oximate; 0.3 M, 10 equiv. per phosphotriester) in dioxane:acetonitrile (1:1, v/v) was added to a fully-protected dimer (14c, 15c or 15d). The reaction mixture was stirred for 36 h at 20°C. Dowex 50 W cation-exchange resin (100-200 mesh, 7 g/mmol oximate) was added. After 5 min, the resin was filtered off and washed with water. The filtrate was concentrated and aqueous ammonia (25%) was added. The reaction vessel was sealed and kept at 50°C for a period of 48 h. The reaction mixture containing dimer 14c or 15c (R^2 = lev) was concentrated to a small volume and analysed by hplc. The reaction mixture containing dimer 15d (R^2 = DMTr) was concentrated and a mixture of acetic acid:water (4:1, v/v) was added. After 1 h at 20°C, the solution was concentrated and coevaporated three times with water. The aqueous layer was washed with chloroform, concentrated to a small volume and analysed by hplc.

In all cases hplc analysis showed the presence of one main product. Crude deprotected dimers were purified by anion-exchange chromatography on a column of DEAE Sephadex A25 (HCO_3^-) according to the same procedure as described before [28]. Digestion of the thus obtained pure dimers with spleen phosphodiesterase [29] gave the expected products in the correct ratio.

REFERENCES

[1] Reese, C.B. (1978) Tetrahedron 34, 3143

[2] Schaller, H., Weimann, G., Lerch, B., and Khorana, H.G. (1963) J. Am. Chem. Soc. 85, 3821

[3] van Boom, J.H. and Burgers, P.M.J. (1976) Tetrahedron Lett., 4875

[4] Chattopadhyaya, J.B., Reese, C.B., and Todd, A.H. (1979) J.C.S. Chem. Comm., 987

[5] Reese, C.B., Titmas, R.C., and Yau, L. (1978) Tetrahedron Lett., 2727

[6] Jones, S.S., Rayner, B., Reese, C.B. Ubasawa, A., and Ubasawa, M. (1980) Tetrahedron 36, 3075

[7] de Rooij, J.F.M., Wille-Hazeleger, G., van Deursen, P.H., Serdijn, J., and van Boom, J.H. (1979) Recl. Trav. Chim. Pays-Bas 98, 537

[8] Arentzen, R., van Boeckel, C.A.A., van der Marel, G., and van Boom, J.H. (1979) Synthesis, 137

[9] Katagiri, N., Itakura, K., and Narang, S.A. (1975) J. Am Chem. Soc. 97, 7332

[10] Chattopadhyaya, J.B. and Reese, C.B. (1979) Tetradedron Lett., 5059

[11] König, W. and Geiger, R. (1970) Chem. Ber. 103, 788

[12] van der Marel, G., van Boeckel, C.A.A., Wille, G., and van Boom, J.H. (1981) Tetrahedron Lett., 3887

[13] van Boeckel, C.A.A., van der Marel, G., Wille, G., and van Boom, J.H. (1981) Chem. Lett., 1725

[14] Owen, G.R., Reese, C.B., Ransom, C.J., van Boom, J.H., and Herscheid, J.D.H. (1974) Synthesis, 704

[15] Ogilvie, K.K. (1973) Can. J. Chem. 53, 3799

[16] Corey, E.J. and Venkateswarlu, A. (1972) J. Am. Chem. Soc. 94, 6190

[17] Crea, R., Kraszewski, A., Hirose, T., and Itakura, K. (1978) Proc. Natl. Acad. Sci. USA 75, 5765

[18] Takaku, K., Yoshida, M., Kamaike, K., and Hata, T. (1981) Chem. Lett., 197

[19] De Bernardini, S., Waldmeier, F., and Tamm, Ch. (1981) Helv. Chim. Acta 64, 2142

[20] Balgobin, N., Josephson, S., and Chattopadhyaya, J.B. (1981) Acta Chem. Scan. B35, 4

[21] Takaku, H., Yoshida, M., Kato, M., and Hata, T. (1979) Chem. Lett., 811

[22] Broka, C., Hozumi, T., Arentzen, R., and Itakura, K. (1980) Nucleic Acids Res. 8, 5461

[23] Pfleiderer, W., Uhlmann, E., Charubala, A., Flockerzi, D., Silber, G., and Varma, R.S. (1980) Nucleic Acids Res. Symp. Series 7, 61

[24] Uhlmann, E. and Pfleiderer, W. (1981) Helv. Chim. Acta 64, 1688

[25] Reese, C.B. and Ubasawa, A. (1980) Tetrahedron Lett., 2265

[26] Reese, C.B. and Zard, L. (1981) Nucleic Acids Res. 9, 4611

[27] Agarwal, K.L. and Riftina, F. (1978) Nucleic Acids Res. 5, 2809

[28] de Rooij, J.F.M., Wille-Hazeleger, G., Burgers, P.M.J., and van Boom, J.H. (1979) Nucleic Acids Res. 6, 2337

[29] van Boom, J.H. and de Rooij, J.F.M. (1977) J. Chromatogr. 131, 165

CHEMICAL SYNTHESIS OF OLIGODEOXYNUCLEOTIDES USING THE PHOSPHITE TRIESTER INTERMEDIATES

Marvin H. Caruthers

Department of Chemistry
University of Colorado

SUMMARY

The principles and experimental details for the synthesis of oligodeoxynucleotides using the phosphite triester intermediates are described. The first nucleoside is attached via it's 3'-hydroxyl function to an activated hplc-grade silica gel. The preparation of the protected and activated nucleotides, the cycle for chain elongation and the consecutive work-up of the completed oligonucleotides are outlined.

1 INTRODUCTION

Oligodeoxynucleotides are currently synthesized using the following procedure. Various sections of this protocol have been published previously. For example, the procedures for preparing succinylated nucleosides [1-3] and for modifying silica gel [3,4] are either published or in press. Additionally, the procedures for synthesis of nucleoside phosphoramidites have also been published [5].

2 EXPERIMENTAL SECTION

2.1 A summary of reagents, solvents and equipment

2.1.1 Reagents and solvents that can be used without purification

Acetic acid; acetic anhydride; ammonium hydroxide; t-butylamine; benzophenone; dicyclohexylcarbodiimide; diisopropylethylamine; dimethylamine; 4-dimethylaminopyridine; ninhydrin spray; p-nitrophenol; succinic anhydride; thiophenol; p-toluenesulfonic acid; p-toluenesulfonyl chloride; trichloroacetic acid; 3-triethoxysilylpropylamine; triethylamine; trimethylsilyl chloride.

Acetonitrile; chloroform; dichloromethane; dimethylformamide; dioxane; ethanol; ether; ethyl acetate; formamide; methanol; nitromethane; pentane; tetrahydrofuran; toluene.

Calcium hydride; citric acid; Fractosil 200 "Merck"; iodine; lithium chloride; phosphorus trichloride; phosphorus pentoxide; silica gel (drying agent); sodium chloride; sodium sulfate; anhydrous zinc bromide (The $ZnBr_2$ sample should, however, be tested for activity. Some commercial samples of $ZnBr_2$ have been found to be zinc oxide and zinc hydroxide.); fluorescent tlc silica gel plates; Sephadex G-50; nitrogen or argon tank; molecular sieves 4 Å.

2.1.2 Reagents and solvents that must be purified or prepared

acetonitrile: reflux and distill from phosphorus pentoxide, reflux over calcium hydride and remove just prior to use in condensation

2.6-lutidine: reflux and distill from calcium hydride, distill from 4 Å molecular sieves

tetrahydrofuran: dry over sodium-benzophenone, distill fresh

pyridine: reflux over p-toluenesulfonyl chloride (60 g per L), distill and store over activated 4 Å molecular sieves

chloroform, acid-free: pass over basic, chromatographic aluminia

tetrahydrofuran:water:lutidine (2:2:1, v/v)

tetrahydrofuran:water:lutidine (2:2:1, v/v) containing 0.2 M iodine. Filter this solution through a millipore filter before use.

dry tetrahydrofuran containing 4-dimethylaminopyridine (6.5%, w/v);
acetic anhydride: 2.6-lutidine (1:1, v/v);
2% trichloroacetic acid in dichloromethane
thiophenol:dioxane:triethylamine (1:2:2, v/v)
hplc elution reagent: 1 M triethylammonium acetate

2.1.3 Apparatus and glassware

Test tubes equipped with teflon lined screw caps (10-12 mL); assortment of glass syringes and metal needles, 1 mL to 5 mL volume; sintered glass funnels, 1 mL to 5 mL volume, medium sinter; dry nitrogen; rubber stoppers; side arm erlenmeyer suction flasks attached to water aspirators; plastic wash bottles; gel electrophoresis apparatus; hplc equipped with an ion exchange column; drying oven set at 50°C or higher; distillation apparatus; addition funnels; round bottom flasks and stirring apparatus.

2.1.4 General methods

Thiophenol, 4-dimethylaminopyridine, anhydrous $ZnBr_2$ and iodine can be purchased from Aldrich or other suppliers and used without further purification. Reagent grade acetic anhydride, triethylamine and t-butylamine are used as received. Common solvents such as tetrahydrofuran, dioxane, acetonitrile, nitromethane and methanol are stored over activated (overnight at 160°C in a well ventilated oven) 4 Å molecular sieves and used without further purification. Dry acetonitrile is obtained by refluxing reagent grade solvent over P_2O_5 for 6-8 h and distilling the constant boiling fraction. This fraction is then refluxed over CaH_2 for 6-8 h and distilled just prior to usage. 2,6-Lutidine is obtained by refluxing reagent grade solvent over CaH_2 for one h followed by distillation from 4 Å molecular sieves. 2,6-Lutidine is stored in the dark. Tetrahydrofuran is dried over sodium wire (a small amount of benzophenone is added to prevent peroxide formation) and used freshly distilled. 1-H-Tetrazole (Aldrich) is sublimed at 110 to 115°C at 0.05 mm Hg prior to use.

All solution transfers involving dry reagents are completed with clean syringes dried in an oven at 50°C. When sintered glass funnels or test tubes are used as reaction flasks, the wash cycle immediately preceeding the condensation step and the condensation step are completed in an atmosphere of dry nitrogen or argon. This is most easily accomplished by passing a dry nitrogen (argon) line through the top of a rubber stopper attached to the top of a test tube or sintered funnel. A second hole in the stopper allows for escape of excess nitrogen.

The stopper is removed during all other steps in the cycle. The main reason for this procedure is to insure that the condensation solution remains dry throughout the synthesis step.

2.2 Individual synthetic steps in the preparation of oligodeoxynucleotides

2.2.1 Synthesis of the support

The initial step is synthesis of the succinylated deoxynucleosides as the p-nitrophenyl esters (Fig. 1, compounds **2a**, **2b**, **2c**, and **2d**). All four are prepared using the same general procedure as outlined in figure 1. To a solution of 5'-dimethoxytritylnucleoside (5 mmol) in anhydrous pyridine is added 4-dimethylaminopyridine (0.61 g; 5 mmol) and succinic anhydride (6.6 g; 6 mmol). The reaction is monitored by tlc (acetonitrile:water, 9:1) and is usually complete after 12 h at 20°C. Occasionally a second portion of succinic anhydride (0.1 g; 1 mmol) is added. The reaction is next quenched with water (0.1 mL) for 10 min at 20°C. The reaction mixture is evaporated *in vacuo*, and then co-evaporated twice with dry toluene (2 x 20 mL). The residue is redissolved in dichloromethane (40 mL) and the solution is washed successively, once with 10% citric acid (10 mL) and twice with water (2 x 10 mL). The organic solution is dried over anhydrous sodium sulfate, and evaporated *in vacuo*. The residue is redissolved in 10 mL dichloromethane (containing 5% pyridine) and the product precipitated into pentane:ether (200 mL; 1:1). The precipitate is dried *in vacuo* (yield: 70-85%). The succinylated nucleoside (1 mmol) is next dissolved in dioxane (4 ml) containing dry pyridine (0.2 mL) and p-nitrophenol (140 mg; 1 mmol) is added. A solution of dicyclohexylcarbodiimide (220 mg; 1 mmol) in anhydrous dioxane (1 mL) is added and the reaction is monitored by tlc (silica gel plate; benzene:dioxane, 3:1). The reaction is virtually complete after 2 h at room temperature. Dicyclohexylurea is removed by centrifugation, and the supernatant containing the desired product is used directly in the condensation reaction.

Deoxynucleosides are attached to the support using the following general procedure. Hplc grade silica gel (12 g, Fractosil 200, "Merck") is exposed to a 15% relative humidity (saturated LiCl solution) for at least 24 h. The silica is then treated with 3-triethoxysilylpropylamine (13.8 g, 60 mmol, 0.01 M in dry toluene) for 12 h at 20°C and 18 h at reflux. It is isolated by centrifugation, washed successively (200 mL, 3 times each) with toluene, methanol and 50% aqueous methanol. The silica is shaken with 50% aqueous methanol (200 mL) at 20°C for 18 h. After isolation by centrifugation, the silica is washed with methanol

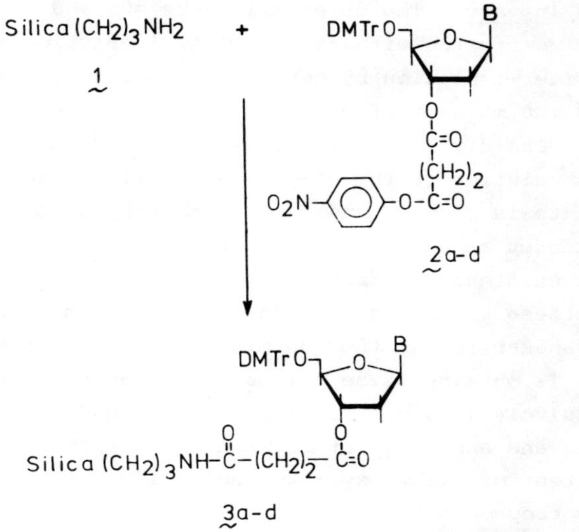

Figure 1 Synthesis of the polymer support. B refers to thymine in 2a and 3a; to N-benzoylcytosine in 2b and 3b; to N-benzoyladenine in 2c and 3c; and to N-isobutyrylguanine in 2d and 3d. DMTr designates the di-p-anisylphenylmethyl protecting group

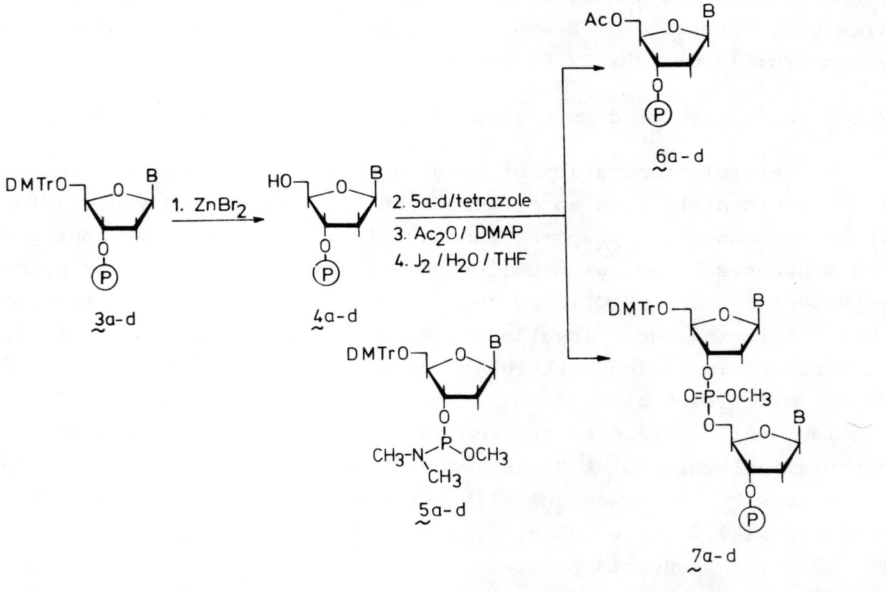

Figure 2 Steps in the synthesis of a dinucleotide. B refers to thymine in 3a, 4a, 5a, 6a, and 7a; to N-benzoylcytosine in 3b, 4b, 5b, 6b, and 7b; to N-benzoyladenine in 3c, 4c, 5c, 6c, and 7c; to N-isobutyrylguanine in 3d, 4d, 5d, 6d, and 7d. DMTr refers to the di-p-anisylphenylmethyl group

and ether and dried *in vacuo*. The dried silica is suspended in 100 mL anhydrous pyridine and treated with trimethylsilyl chloride (15 mL) for 12 h at 20°C. After isolation by centrifugation, the silica is washed 4 times with 200 mL each of methanol, twice with ether, and then dried *in vacuo*. The dry silica (3 g) is suspended in dimethylformamide (5 mL) and a solution of the 5'-dimethoxytritylnucleoside-3'-p-nitrophenylsuccinate in dioxane and 1 mL of triethylamine is added. The suspension is shaken at 20°C for 4 h. A ninhydrin test at this stage indicates the existence of free amino groups on the silica gel. To acylate, "cap", these groups, acetic anhydride (0.6 mL) is added and the mixture is shaken for another 30 min, after which time a negative ninhydrin test is obtained. The silica is isolated by centrifugation, washed successively (100 mL, 3 times each) with dimethylformamide, 95% ethanol, dioxane and ethyl ether, and then dried *in vacuo*. Analysis for the extent of dimethoxytritylnucleoside attached to the support is done spectrophotometrically. An accurately weighed sample of silica (10-15 mg) is treated with 5 mL 0.1 M p-toluenesulfonic acid in acetonitrile and the optical density of the supernatant obtained after centrifugation is measured at 498 nm. (The extinction coefficient of an acid solution of dimethoxytritanol is 70[$cm^2 \cdot \mu mol^{-1}$]. A typical preparation leads to the following amounts of nucleosides bound to silica gel: $(MeO)_2$TrdT, 62 μmol/g; $(MeO)_2$TribdG, 56 μmol/g; $(MeO)_2$TrbzdA, 65 μmol/g; $(MeO)_2$TrbzdC, 68 μmol/g.

2.2.2 Synthesis of deoxynucleoside phosphoramidites

The careful preparation of compounds **5a-d** is of critical importance for the preparation of an oligodeoxynucleotide in high yield (Fig. 2). These compounds are prepared essentially as described previously [5]. The synthesis begins with the preparation of chloro-N,N-dimethylaminomethoxyphosphine ($CH_3OP(Cl)N(CH_3)_2$) which is used as a monofunctional phosphitylating agent (synthesis, see Seliger et al.). A 250 mL addition funnel is filled with 100 mL of precooled (-78°C) anhydrous dimethylamine (45.9 g, 1.02 mol). The addition funnel is wrapped with aluminum foil containing dry ice in order to avoid evaporation of dimethylamine. This solution is added dropwise at -15°C (ice-acetone bath) over 2 h to a mechanically stirred solution of methoxydichlorophosphine (47.7 mL, 67,32 g, 0.51 mol) in 300 mL of anhydrous ether. The addition funnel is removed and the 1 L, three-necked round bottom flask is stoppered with serum caps tightened with copper wire. The suspension is mechanically stirred for 2 h at room temperature. The suspension is filtered and the amine hydrochloride salt is washed with 500 mL anhydrous ether. The filtrate and washings are combined and ether is distilled at atmospheric pressure. The residue is distilled under reduced pressure. The product is collected at 40-42°C at 13 mm Hg and is isolated in 71% yield (51.1 g, 0.36 mol). d^{25} = 1.115 g/mL. ^{31}P-nmr, δ = -179,5 ppm ($CDCl_3$) with respect to internal 5% v/v aqueous H_3PO_4 standard. H-nmr doublet at 3.8 and 3.6 ppm J_{P-H} = 14 Hz (3 H, OCH_3) and two singlets at 2.8 and 2.6 ppm (6 H, $N(CH_3)_2$). The mass spectrum shows a parent peak at m/e = 141.

Compounds 5a-d (Fig. 2) are prepared by the following procedure: 5'-O-di-p-anisylphenylmethylnucleoside (1 mmol) is dissolved in 3 mL of dry, acid-free chloroform and diisopropylethylamine (516 mg, 4 mmol) in a 10 mL Erlenmeyer flask preflushed with dry nitrogen. $CH_3OP(Cl)N(CH_3)_2$ (445 mg, 2 mmol) is added dropwise (30-60 sec) by syringe to the solution under nitrogen at room temperature. After 15 min the solution is transferred with 35 mL of ethyl acetate into a 125 mL separatory funnel. The solution is extracted four times with an aqueous, saturated solution of NaCl (80 mL). The organic phase is dried over anhydrous Na_2SO_4 and evaporated to a foam under reduced pressure. The foam is dissolved with toluene (10 mL) (5d is dissolved with 10 mL of ethyl acetate) and the solution is added dropwise to 50 mL of cold hexane (-78°C) with vigorous stirring. The cold suspension is filtered and the white powder is washed with 75 mL of cold hexane (-78°C). The white powder is dried under reduced pressure in a dessicator and stored under nitrogen. Isolated yields of compounds 5a-d are 90-94%. The purity of the products is checked by ^{31}P-nmr. Compounds 5a-d are characterized as two peaks between -146 and -145 ppm. Various impurities are sometimes observed with peaks between 0 and +10 relative to phosphoric acid. These impurities do not appear to inhibit the condensation reactions.

2.2.3 Outline of the synthesis cycle

The appropriately derivatized deoxynucleosides attached covalently to silica gel (compound 3a, 3b, 3c or 3d) are treated with 1.5 mL of a saturated solution of anhydrous $ZnBr_2$ in nitromethane:methanol (95:5) for 4 min. The support is then washed with 2 mL nitromethane followed by 2 mL methanol. Before the condensation step the silica gel is carefully washed several times with 2 mL dry acetonitrile under a dry inert atmosphere (N_2). Stock solutions of sublimed tetrazole and appropriately protected 2'-deoxynucleoside-3'-N,N-dimethylaminomethoxyphosphines are prepared in dry acetonitrile and stored over an inert gas atmosphere (N_2). Usually these stock solutions are sufficient for at least three condensations. For each µmol of deoxynucleoside attached covalently to silica gel, tetrazole (60 µmol) and the deoxynucleoside-3'-

Table 1 A summary of the chemical steps for one synthetic cycle

reagent or solvent	purpose	time (min)
satd. $ZnBr_2$ in 5% methanol/nitromethane	detritylation	4
nitromethane	wash	1
methanol	wash	1
acetonitrile	wash	1
activated nucleotide in acetonitrile	add one nucleotide	5
capping solution	acylation	2
lutidine:tetrahydrofuran:H_2O (1:2:2)	wash	2
I_2 solution	oxidation	2
acetonitrile	wash	2
nitromethane	wash	2

N,N-dimethylaminomethoxyphosphine (20 μmol) are 0.1 M in the condensation mixture (0.6 mL and 0.2 mL). The condensation reaction is completed under an inert gas atmosphere (N_2) and stopped after 5 min. The excess phosphine can be reduced to 5 or 10 fold over support bound nucleoside if rigorously dry reaction conditions are used. This is possible in a closed, machine type device (see Matteucci and Caruthers [4]). However, if test tubes or sintered funnels are used as reaction vessels, a larger excess of phosphine should be used in order to insure that conditions are anhydrous. Immediately following the condensation reaction, the silica gel is washed with 1-2 mL of a tetrahydrofuran:water:2,6-lutidine solution (2:2:1) for 3 min. Oxidation of trivalent phosphorus to pentavalent phosphate is with 1.5 mL of iodine in tetrahydrofuran:water:2,6-lutidine (2:2:1) for 5 min. The silica gel is washed with 2 mL methanol followed by 2 mL tetrahydrofuran. Acylation of unreactive hydroxyl groups is completed by adding first a solution of 4-dimethylaminopyridine in dry tetrahydrofuran (2 mL; 6.5% w/v) and then a solution of acetic anhydride in 2,6-lutidine (0.4 mL; 1:1) to the support ("capping"). After 5 min, this acylation solution is removed and the silica gel is washed with 2 mL each of methanol and nitromethane. This step completes one synthesis cycle.

2.3 Isolation of synthetic oligodeoxynucleotides

After completion of the appropriate synthesis cycle, oligodeoxynucleotides free of protecting groups and side products are isolated using the following procedure. Silica gel containing the oligodeoxynucleotide is first treated with 5 mL of a solution containing thiophenol:triethylamine (1:2:2) for 90 min at room temperature followed by a wash with 1.5 mL methanol and 2 mL ether. This deprotection step removes the methyl phosphotriester protecting group. The support is next treated with 3 mL concentrated ammonium hydroxide for 24 h at 60°C. The liquid phase is evaporated to dryness in vacuo and the residue is treated with a 2 mL solution of t-butylamine in methanol (1:1) for 24 h at 60°C in a screw cap vial. The reaction mixture is evaporated to dryness in vacuo and then loaded on a Sephadex G-50 column (1 cm x 50 cm). The column is eluted with 10 mM Tris-HCl, pH 7.5, 5 mM EDTA and 20 mM NaCl. Fractions containing oligodeoxynucleotides are pooled and the DNA is isolated by ethanol precipitation. The crude DNA pellet is dissolved in a minimal volume of formamide and loaded on a 20% denaturing polyacrylamide gel. After electrophoresis, the bands containing oligodeoxynucleotide products on the gel are visualized using an ultraviolet lamp. The absorbance can be enhanced by placing a silica gel thin layer plate containing a fluorescent dye behind the gel. The gel slice containing the product is eluted and desalted using standard procedures. Usually the product is the major uv light absorbing band on the gel (for page purification of oligonucleotides see Gait, this volume).

ACKNOWLEDGEMENTS

This research was supported by grants from the National Institutes of Health (GM21120 and GM25680). M.H.C. also acknowledges support derived from an NIH Research Career Development Award (1 KO4 GM00076). Several excellent and imaginative graduate and postdoctoral students contributed immeasureably to the development of this chemical methodology. Their names appear in the references cited.

REFERENCES

[1] Gait, M.J. (1980) Nucleic Acids Research 8, 1081-1096

[2] Chow, F., Kempe, T., and Palm, G. (1981) Nucleic Acids Research 9, 2807-2817

[3] Caruthers, M.H. (1982) Proceedings of the Symposium on Promoters Structure and Function, in press

[4] Matteucci, M.D. and Caruthers, M.H. (1981) Journal of the American Chemical Society 103, 3185-3191

[5] Beaucage, S.L. and Caruthers, M.H. (1981) Tetrahedron Letters 22, 1859-1862

SOLID-PHASE SYNTHESIS OF OLIGONUCLEOTIDES USING THE PHOSPHITE METHOD

Hartmut Seliger, Sonja Klein, Chander K. Narang, Barbara Seemann-Preising, Josef Eiband, and Norbert Hauel

Sektion Polymere
Universität Ulm

SUMMARY

The experiment describes the synthesis of the hexanucleotide dCCGAGG and dCCTCGG on a polymer support according to the procedure of M.H. Caruthers and coworkers with some modifications. In the first stage 5'-dimethoxytrityl-nucleoside-3'-methoxydimethylamino-phosphoramidite is prepared. Furthermore the coupling to the polymer support, the reaction cycle for chain elongation, the release of the prepared oligonucleotides from the support, their deprotection and purification are explained.

1 INTRODUCTION

Synthesis on polymeric carriers has been a major simplification in multistep preparations of biopolymers of defined sequence. The essential feature of this method is the use of a macromolecular blocking group, which serves a double purpose, namely

a) to protect one of the nucleotide chain termini,
b) to provide for a specific and efficient separation of the thus protected oligonucleotide chains from other components of the reaction mixture.

In most cases insoluble synthetic resins serve as macromolecular supports. The solid phase synthesis, therefore, begins with the covalent attachment of the initial nucleoside or nucleotide monomer to the carrier matrix. Then, the oligonucleotide chains are elongated by stepwise condensation of the immobilised molecules with further nucleotide monomers, which, in activated state, are applied in solution, and thus are mobile reaction partners. In addition to the condensation steps all other reactions, i.e. blocking/deblocking and oxidation, are done heterogeneously. Each reaction step is followed by washing and filtration procedures to remove the mobile reactants. For the addition of each single nucleotide unit a series of consecutive reaction and washing/filtration steps, termed "reaction cycle", has to be followed. Finally, when the desired sequence is completed, the immobilised material is cleaved from the carrier and the end product is purified by conventional techniques.

This concept for multistep syntheses, originally developed by R.B. Merrifield [1] and R.L. Letsinger [2], has the advantage, that separation steps, in the course of the chain elongations, are reduced to simple and time-saving washing/filtration operations, thus avoiding laborious chromatographic techniques. Since all reactions and work-ups are done heterogeneously, there are no complications due to differences in solubility of the components. The dissolved reaction partners can be applied in ample excess, and the reaction can easily be rerun in order to increase the yields. Altogether this enables a certain standardisation of reaction and work-up conditions. Therefore, solid-phase procedures are the basis for an automation of polynucleotide synthesis.

A major disadvantage of the carrier concept is the fact that the polymer support is a polyfunctional protecting group. A great number of oligonucleotide chains are attached to one resin particle. Ideally, they should have to grow simultaneously, i.e. all reactions should have to proceed quantitatively. Neither in peptide nor in oligonucleotide syntheses can this goal be completely achieved. Therefore, the efficiency of techniques used to separate the resulting mixture of homologues and possible failure sequences is the main restriction on the length of chains to be synthesised on the support.

A further complication of support reactions is that they proceed within a "polymer phase" ("solid phase" according to Merrifield, which restricts the free movement of solvent and solute molecules. Reactions with the immobilised partners may therefore be retarded or even controlled by the diffusion of the mobile reactants. Steric and neighbouring group effects of the matrix may either enhance or suppress side reactions. In addition, problems related to the solvation of the polymer matrix, the immobilised oligonucleotide chains and the mobile reactants, as well as adsorption problems, may influence the reactions on solid supports.

Such effects and interactions have to be considered when choosing the proper support matrix, the type of anchor for the oligonucleotide chains, and the strategy for the oligonucleotide synthesis itself. A great number of support systems have been proposed for different variants of internucleotide bond formation [3], although, the results have been relatively poor for more than a decade. During the last few years the interest in solid phase oligonucleotide synthesis has dramatically increased. New types of supports, and improvements in the purification of the products from the solid phase synthesis by hplc and affinity chromatography have made carrier methods more attractive for phosphodiester routes [4]. Synthesis strategies leading to blocked internucleotide bonds (phosphotriester route) seem to be even more suited to preparations on polymeric carriers, probably due to a better compatibility of neutral oligonucleotide chains and rather hydrophobic matrices in organic solvents. The success of phosphotriester methods [5] in the synthesis of genes has stimulated the development of solid phase methods adopting this strategy [6]. During recent years an alternative route for internucleotide bond formation has been developed, referred to as the "phosphite method" [7].

It features the use of highly reactive phosphorous acid monoester dichloridites as reagents for "preformation" of the internucleotide bond as a protected dinucleoside phosphite linkage. The latter is oxidised to a phosphate bond after each elongation cycle. Since the high reactivity of the phosphite intermediates impairs the selectivity of the internucleotide bond formation, it is almost essential, in this case, to keep the growing chains locked to a polymer support. Thus, the addition of solid phase techniques has been a breakthrough for the phosphite approach, making it currently the fastest and probably most efficient method for oligonucleotide synthesis. Procedures from several laboratories have been published recently and reports on the automation of this process have been made [8-10].

In the experiments described we use the procedure developed in the laboratory of M.H. Caruthers [9,11,12], together with some modifications recently developed by our group*.

2 EXPERIMENTAL SECTION

2.1 Synthesis of the hexanucleotides dCCGAGG and dCCTCGG on a polymer support

2.1.1 Preparation of 5'-dimethoxytrityl-nucleoside-3'-methoxydimethyl-amino-phosphoramidites

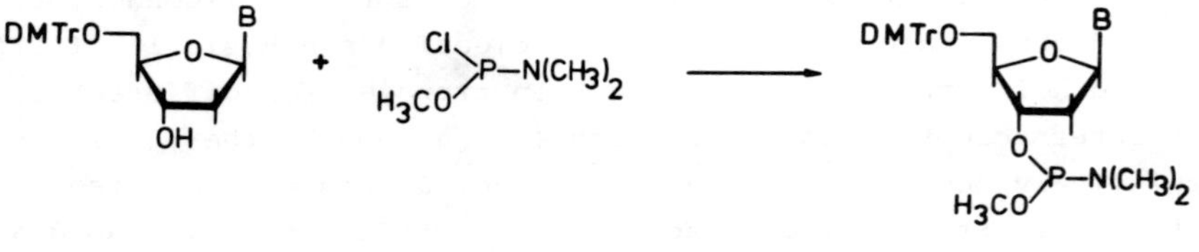

DMTrO–(nucleoside, B, 3'-OH) + $Cl(H_3CO)P{-}N(CH_3)_2$ ⟶ DMTrO–(nucleoside, B, 3'-O–$P(OCH_3){-}N(CH_3)_2$)

1: B = N-6-benzoyladenine
2: B = N-4-benzoylcytosine
3: B = N-2-isobutyrylguanine
4: B = thymine
DMTr: 4,4'-dimethoxytrityl

Figure 1 Synthesis of deoxynucleoside phosphoramidites

Reagents:

1.0 mmol DMTrdN (N = bzA, bzC, ibG, T)
260 µL $(CH_3O)P(Cl)N(CH_3)_2$
prepared according to appendix 1
600 mg diisopropylethylamine
abs. $CHCl_3$; NaCl-solution, saturated; ethyl acetate; toluene; hexane

Procedure:

The protected nucleoside is dissolved in 3 mL abs. $CHCl_3$, and 600 mg (0.8 mL) of dry diisopropylethylamine is added. To this solution the phosphochloridite is added dropwise within 1 min with stirring, in a dry box. The mixture reacts at room temperature within 12

---

*We are greatly indebted to M.H. Caruthers as well as to E.L. Winnacker and T. Dörper for helpful advice and valuable information.

min. The solution is diluted with 35 mL ethyl acetate and extracted with 20 mL saturated NaCl solution. This extraction is repeated 4 times with 20 mL of saturated NaCl each. The organic phase is dried over $MgSO_4$, concentrated in a rotary evaporator using a water pump and finally under high vacuum. The residue is dissolved in 10 mL toluene and added dropwise to 50 mL cold n-hexane (-78°C). The precipitate is immediately washed with 75 mL cold n-hexane and dried in vacuo.

A sample is subjected to ^1H-nmr spectroscopy in $CDCl_3$. The addition of the phosphite residue is indicated by two doublets centered at 3.4 ppm tetramethylsilan (TMS).

2.1.2 Oligonucleotide synthesis on the polymer support - The reaction cycle for chain elongation

Reagents:

100	mg	polymer support, containing ca. 10 µmol immobilised DMTrdibG, prepared according to lit. 9 (figure 2)
0.1	mmol	DMTrdN-OP$(OCH_3)N(CH_3)_2$, prepared as described in app. 1
35	mg	(0.5 mmol) tetrazol
		$ZnBr_2$, solution in nitromethane (70 g $ZnBr_2$ in 500 mL nitromethane + 5 mL H_2O [10])
250	mg	iodine
2.4	g	dimethylaminopyridine; acetonitrile, abs.; acetonitrile, abs., passed over Alox (Woelm B, act. I); 0.1 M p-toluene sulfonic acid in acetonitrile; n-butanol; lutidine; THF; acetonitrile; acetic anhydride; collidine; nitromethane

Procedure:

Before starting the experiment read the suggestions as summarised in appendix 2.

A. Deblocking: 100 mg support material are incubated for 5-10 min at room temperature with $ZnBr_2$ solution. The $ZnBr_2$ solution is removed by suction, and the resin is further washed three times with $ZnBr_2$ solution (the last wash should be colorless).

B. Washing: 4 times with ca 2 mL each of n-butanol:lutidine:THF (4:1:5, v/v).

C. Drying: Ca 15 mL abs. acetonitrile are passed through the resin bed; the resin is further dried by removing residual solvent under high vacuum.

D. Condensation: DMTrdN-OP$(OCH_3)N(CH_3)_2$, 10fold excess over immobilised nucleotide (oligonucleotide, ca 100 µmol), and 35 mg tetrazole,

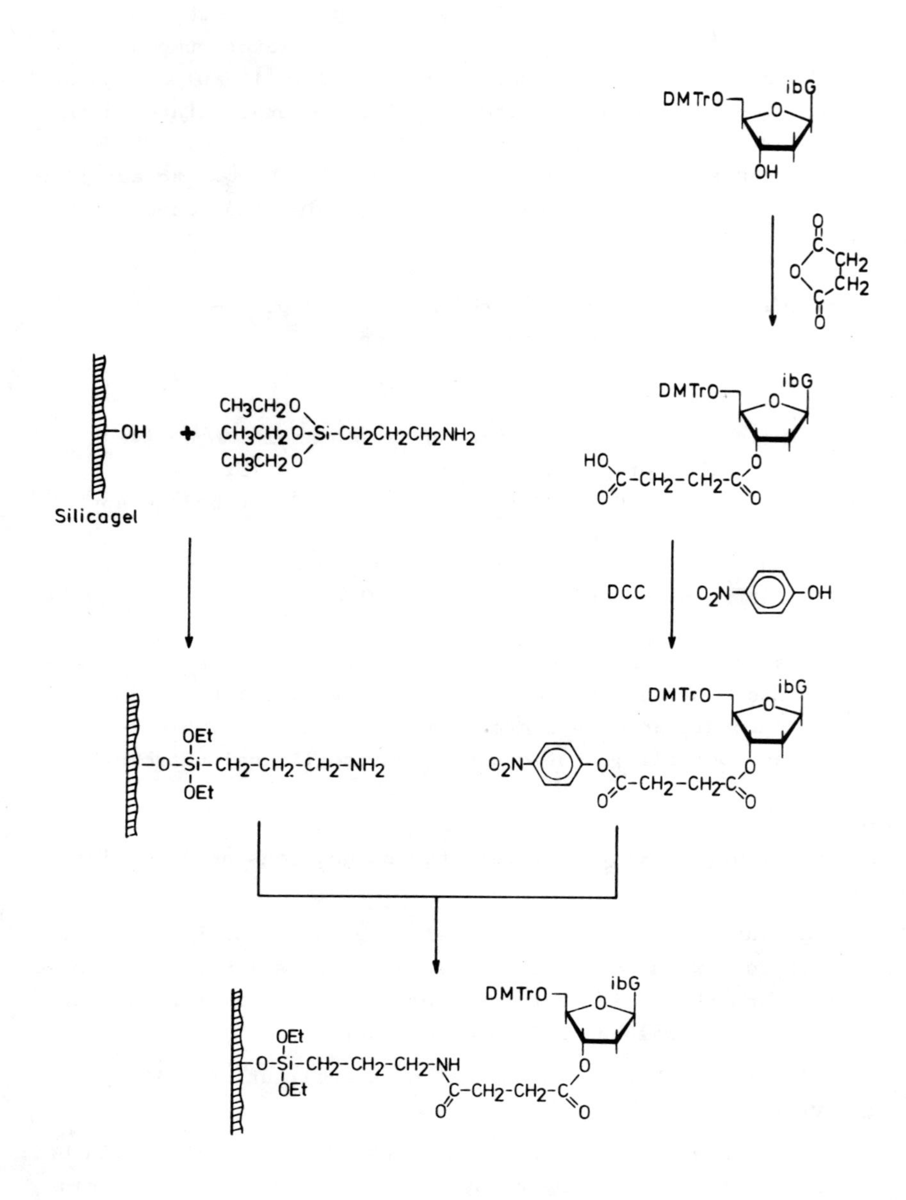

Figure 2 Preparation of polymer support

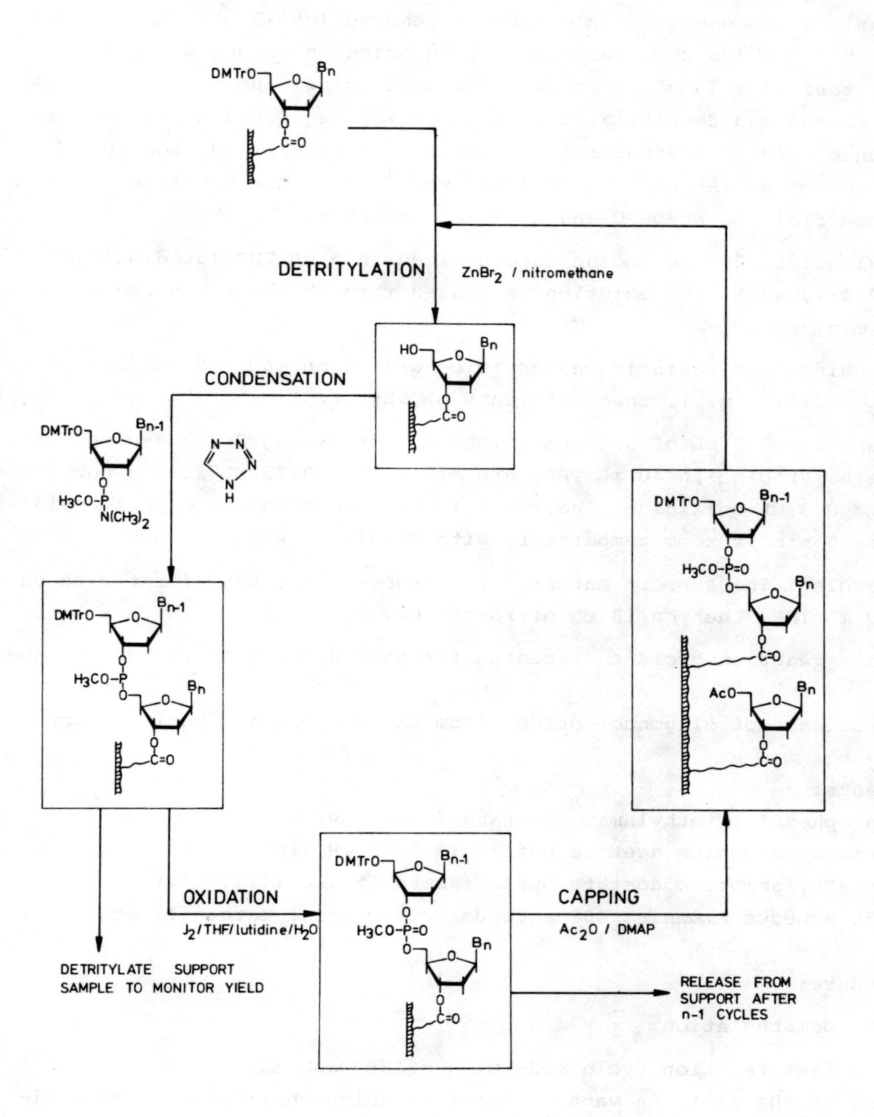

Figure 3 Oligonucleotide synthesis on polymer support - reaction cycle for chain elongation

dissolved in 4 mL abs. acetonitrile, are added to the resin, and the suspension is shaken for 45 min.

E. Washing, drying, monitoring: The resin is washed with acetonitrile until all monomeric nucleotide is removed (ca 15 mL), then 3 times with ether (ca 2 mL per wash), then dried in vacuo. A small sample of resin (ca 1 mg) is removed for monitoring, the exact weight determined and detritylated in a given volume of 0.1 M toluene sulfonic acid in acetonitrile. The yield is determined from the absorption at 498 nm (ε = 70 [$cm^2 \cdot \mu mol^{-1}$]). If the coupling yield is insufficient, steps D and E can be repeated.

F. Oxidation: 250 mg iodine is dissolved in 5 mL THF:lutidine:H_2O (2:1:1, v/v). The solution is passed through the resin bed within 5 min.

G. Washing: The resin is washed twice with 8 mL each of THF:lutidine:H_2O (2:1:1, v/v), then with ca 5 mL THF.

H. Capping: 2.5 mL of a stock solution, containing 2.4 g 4-dimethylaminopyridine in 40 mL THF, are mixed with 0.25 mL acetic anhydride and 0.3 mL collidine. The resin is treated with the capping solution for 5 min at room temperature with gently shaking.

I. Washing: The support material is washed thoroughly, first with ca 10 mL THF, then ca 10 mL nitromethane.

This reaction cycle is repeated for each nucleotide to be attached.

2.2 Release of oligonucleotides from the support and purification

Reagents:

thiophenol:triethylamine:dioxane (1:1:2, v/v)
triethylammonium acetate buffer (0.1 M, pH 7.0)
triethylammonium acetate buffer:acetonitrile (4:1, v/v)
25% aqueous ammonia; 80% aqueous acetic acid; methanol; ether

Procedure:

2.2.1 Demethylation

The last reaction cycle ends with a THF wash (step G above) and drying of the resin in vacuo. Then 6 mL thiophenol:triethylamine:dioxane (1:1:2, v/v) are added and the suspension is shaken for 45 min. The resin is washed with 8 mL each of methanol and ether, then dried in vacuo.

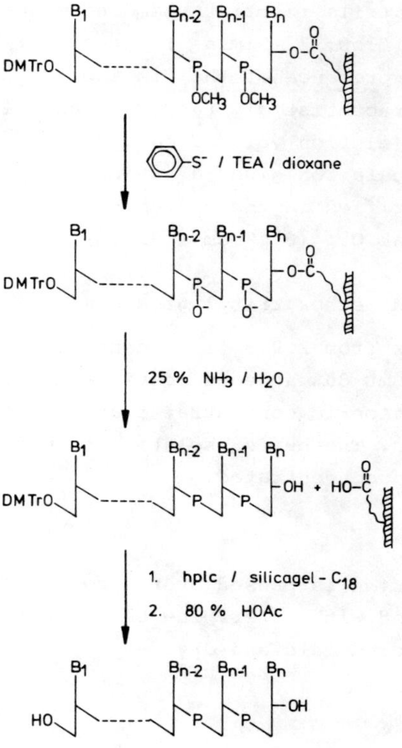

Figure 4 Individual steps in the release of the oligonucleotide from the solid support and the removal of the protective groups

2.2.2 Release of oligonucleotides from the support and deprotection of nucleobases

The support material is suspended in 6 mL of 25% aqueous ammonia and kept at room temperature overnight, then at 60°C for 5 h. The ammoniacal solution is carefully decanted from the resin and concentrated in vacuo.

2.2.3 Liquid chromatography

The residue from 2.2.2 is dissolved in a small volume (ca 1 mL) of triethylammonium acetate buffer, pH 7. A sample is injected into the analytical hplc column: μ-Bondapak C_{18} (Waters), 0.42 cm x 25 cm; eluant: 25% acetonitrile in triethylammonium acetate buffer, pH 7; pressure: 176 bar; flow: 2 mL•min^{-1}; detection: uv-absorption 254 and 280 nm. Due to the presence of the dimethoxytrityl group the product peak should be characteristicly retarded and well resolved from the other by-products (elution volume: ca 8 mL).

If the pilot separation gives a satisfactory result, the bulk of the material is purified in the same manner using a "semi-preparative" column of μ-Bondapak C_{18} (0.78 cm x 25 cm).

2.2.4 Removal of dimethoxytrityl groups

The product peak from 2.2.3 is concentrated to a gum, and the residue treated with 2 mL 80% aqueous acetic acid for 15 min at room temperature. The solution is concentrated and diluted with triethylammonium acetate buffer. The buffer solution is extracted twice with 5 mL ether each time and concentrated.

2.2.5 Desalting

The buffer solution is passed over a column (2 cm x 50 cm) of Biogel P2 and eluted with water. The eluate is lyophilised and the oligonucleotide stored cold and dry.

REFERENCES

[1] Merrifield, R.B. (1963) J. Amer. Chem. Soc. 85, 2149

[2] Letsinger, R.L. and Kornet, M.J. (1963) J. Amer. Chem. Soc. 85, 3045

[3] Kössel, H. and Seliger, H. (1975) Recent Advances in Polynucleotide Synthesis, in: Progress of the Chemistry of Organic Natural Products (W. Hertz, H. Grisebach, and G.W. Kirby, eds.) 32, 297

[4] e.g.: Atherton, E., Gait, M.J., Sheppard, R.C., and Williams, B.J. (1979) Bioorganic Chem. 8, 351;
Potapov, V.K., Veiko, V.P., Koroleva, O.N., and Shabarova, Z.A. (1979) Nucl. Acids Res. 6, 2041;
Narang, C.K., Brunfeldt, K., and Norris, K.E. (1977) Tetrahedron Lett., 1819

[5] Narang, S.A., Brousseau, R., Hsiung, H.M., and Michniewicz, J.J. (1980) Methods Enzymol. 65, 610;
Hsiung, H.M. and Brousseau, R. (1979) Methods Enzymol. 68, 90; and literature cited there

[6] Miyoshi, K. and Itakura, K. (1979) Tetrahedron Lett., 3635;
Itakura, K. et al. (1980) Nucl. Acids Res. 8, 5473, 5491, 5507;
Dembek, P., Miyoshi, K., and Itakura, K. (1981) J. Amer. Chem. Soc. 103, 706;
Crea, R. and Horn, T. (1980) Nucl. Acids Res. 8, 2338;
Duckworth, M.L., Gait, M.J., Goelet, P., Horn, G.F., Singh, M., and Titmas, R.C. (1981) Nucl. Acids Res. 9, 1691;
Shabarova, Z.A. (1980) Nucl. Acids Res. Symp. Ser. 7, 259;
Norris, K.E., Norris, F., and Brunfeldt, K. (1980) Nucl Acids Res. Symp. Ser. 7, 233;
Seliger, H. and Görtz, H.H. (1981) Angew. Chem. 93, 709, Angew. Chem. Intern. Ed. Engl. 20, 683;
Edge, M.D., Greene, A.R., Heathcliffe, G.R., Meacock, P.A., Schuch, W., Scanlon, D.B., Atkinson, T.C., Newton, C.R., and Markham, A.F. (1981) Nature 292, 756

[7] a) Letsinger, R.L. and Lunsford, W.B. (1976) J. Amer. Chem. Soc. 98, 3655;
b) Ogilvie, K.K., Theriault, N., and Sadana, K.L. (1977) J. Amer. Chem. Soc. 99, 7741;
Ogilvie, K.K. and Nemer, M.J. (1981) Tetrahedron Lett., 2531
c) Matteucci, M.D. and Caruthers, M.H. (1980) Tetrahedron Lett., 719
d) Fourrey, J.L. and Shire, D.J. (1981) Tetrahedron Lett., 729
e) Molko, D., Derbyshire, R.B., Guy, A., Roget, A., and Teoule, R. (1980) Tetrahedron Lett., 2159

[8] Alvarado-Urbina, G., Sathe, G.M., Liu, W.-C., Gillen, M.F., Duck, P.D., Bender, R., and Ogilvie, K.K. (1981) Science 214, 270

[9] Matteucci, M.D. and Caruthers, M.H. (1981) J. Amer. Chem. Soc. 103, 3185

[10] Chow, R., Kempe, T., and Palm, G. (1981) Nucl. Acids Res. 9, 2807

[11] Caruthers, M.H., Beaucage, S.L., Efcavitch, J.W., Fisher, E.F., Matteucci, M.D., and Stabinski, Y. (1980) Nucl. Acids Res. Symp. Ser. 7, 215

[12] Beaucage, S.L. and Caruthers, M.H. (1981) Tetrahedron Lett. 22, 1859

Appendix 1: Preparation of methoxy-N,N-dimethylamino-chlorophosphine

1. Synthesis of methoxydichlorophosphine

$$PCl_3 + CH_3OH \longrightarrow CH_3O\text{-}PCl_2 + HCl$$

The preparation follows the procedure given by J.E. Malowan, D.R. Martin and P.I. Pizzolato (1954) Inorganic Synthesis 4, 63 with some modifications as approved during the EMBO course.

Reagents:

275 g, (2.0 mol, 111 mL) PCl_3

64 g, (2.0 mol, 81 mL) dry CH_3OH

Apparatus

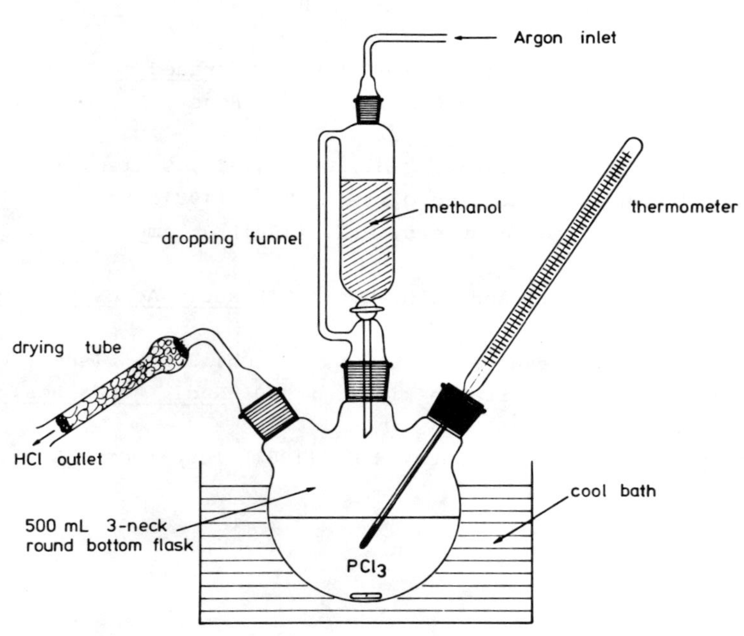

Figure 5 Apparatus for the preparation of methoxydichlorophosphine or methoxy-N,N-dimethylamino-chlorophosphine

- purge apparatus with argon before introducing components
- the use of teflon sleeves for the joints is recommended.

Procedure:

275 g (2.0 mol) PCl_3 are filled into a 3-neck round bottom flask equipped with a magnetic stirrer, dropping funnel and thermometer as shown in the figure above. The PCl_3 is cooled to -10° to -20°C (control temperature with thermometer). Then 64 g (2.0 mol) methanol are added dropwise with stirring such that the temperature does not exceed -10°C (ca 3 h). The mixture is allowed to warm up to 30°C. Then a slow stream of argon is passed through the solution overnight (replace thermometer by argon inlet). The escaped gas is checked for the pH-value to make sure that the formation of HCl has ceased.

The reaction mixture is cooled in an ice bath and residual HCl is removed by applying a vacuum of 100 Torr through the drying tube.

The flask is again allowed to warm up to 30°C and residual PCl_3 is distilled at 30°C, 100 Torr. The vacuum is removed by introducing argon and the distillation is continued under normal pressure. For heating the use of a silicon oil bath is recommended the temperature of which is controlled by a contact thermometer.

The bath temperature should not be raised above 120°C. The boiling point of the crude distillate rises slowly from 83-90°C. Several fractions may be taken, but one should not attempt to obtain high purity material in this first distillation. Rather a fairly steady and rapid flow of the distillate should be accomplished taking care not to heat the content of the flask too long and to a too high temperature.

A yellow precipitate may form on continued distillation, which may lead to instanteneous decomposition of the residual material in the flask with inflammation. Although yields of up to 50% $(CH_3O)PCl_2$ may be obtained by this procedure, as reported in the literature, it is highly recommended to discontinue the distillation, when substantial quantities of yellow precipitate start to form.

The material remaining in the flask after the distillation may be decomposed by dropwise addition of a butanol:chloroform (1:1, v/v) mixture with cooling in dry ice/aceton and subsequent treatment with aqueous sodium hydroxide (2 M).

The product from the first distillation is twice redistilled under normal pressure using a fractionation column (e.g. a 30 cm Vigreeux column). The second distillation yields material free from PCl_3 boiling constantly at 92-93°C, 760 Torr. The purity may be checked by ^{31}P-nmr [12].

2. Synthesis of methoxy-N,N-dimethylamino-chlorophosphine

$$Cl_2P{-}OCH_3 + (CH_3)_3Si{-}N(CH_3)_2 \longrightarrow (CH_3)_2N{-}P(Cl){-}OCH_3 + (CH_3)_3Si{-}Cl$$

Reagents:

17 g (128 mmol, 12 mL) $(CH_3O)PCl_2$

20 mL (127 mmol) $(CH_3)_2N{-}Si(CH_3)_3$ (Fluka)

90 mL ether abs.

Apparatus

As shown in figure 5.

Procedure

After purging with argon, 60 mL abs. ether are filled into the reaction flask and cooled to -20°C. Then 12 mL $(CH_3O)PCl_2$ are introduced with stirring. To this solution 20 mL trimethylsilyl-N,N-dimethylamine dissolved in 30 mL ether are added dropwise with stirring, while a slight stream of argon is maintained through the apparatus. The temperature of the reaction mixture should be controlled with the thermometer and not exceed -15°C. A white precipitate forms during the reaction. The reaction mixture is allowed to warm up to room temperature and is further stirred overnight. The product is distilled in vacuo using a Vigreux column (30 cm) under argon. The product boiling at 59-60°C/30 Torr is collected and checked for purity by ^{1}H-nmr. The yield, based on $(CH_3O)PCl_2$ is generally more than 80%. The material should be stored under argon and protected from moisture.

Spontaneous decomposition and inflammation has also been reported for this compound although the hazard seems to be lower for $(CH_3O)PCl(N(CH_3)_2)$ as well as for purified fractions of $(CH_3O)PCl_2$

Appendix 2: Apparatus for oligonucleotide synthesis (for preparations on a 100–200 mg gel scale)

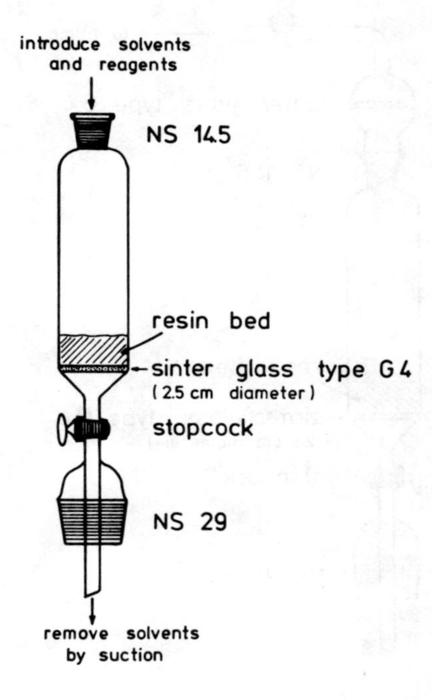

Figure 6 Schematic drawing of the apparatus

Suggestions for handling the device during the reactions cycles (see 2.1)

- apply argon to funnel before introducing the nucleoside phosphoramidite. All other steps can be done without protection from air.
- for drying in high vacuum apply sintered glass adaptor on top of the funnel to prevent resin from blowing out (figure 2)

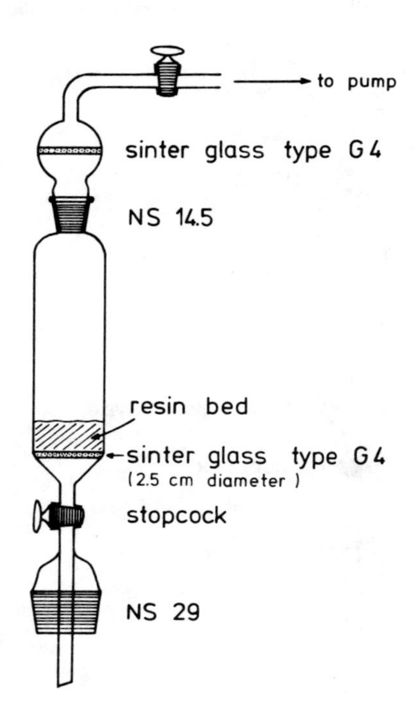

Figure 7 Top adaptor for reaction vessel

- after condensation (step D) and acetonitrile wash use a simple spot test to check for complete removal of nucleoside phosphoramidite: Apply a drop of $ZnBr_2$:nitromethan solution to a filter paper, dry the spot and apply a drop of the final acetonitrile wash solution on the same spot. If red color develops, continue washing with acetonitrile. Incomplete washing may simulate too high yields in the monitoring step.
- gentle shaking or rocking of resin suspension inside the funnel is recommended to speed up diffusion of dissolved reactants. Stirring is not recommended since it leads to mechanical breakage of the resin and consequently clogging of the filter.
- rinse walls of funnel, when applying solvents to wash down resin particles clinging to glass walls.
- solvents, except for acetonitrile used in step D should be high standard of purity but must not be necessarily dry. All solvents used should be free of acid.
- for stop cocks use high vacuum grease, which is insoluble in acetonitrile (e.g. Silicon high vacuum grease from E. Merck, Darmstadt).

SOLID-PHASE SYNTHESIS OF OLIGONUCLEOTIDES USING THE PHOSPHORAMIDITE METHOD

Ernst-L. Winnacker and Thomas Dörper

Institut für Biochemie der
Ludwig-Maximilians-Universität, München

SUMMARY

The experiment describes the synthesis of oligonucleotides using deoxynucleosidephosphoramidites as condensing agents. In the first step these reagents are prepared according to the procedure of Caruthers [1]. In addition, various chemical steps making up the reaction cycle for the chain elongation are outlined.

1. INTRODUCTION

The solid phase synthesis begins with the covalent attachment of the initial 3'-deoxynucleosides to the carrier matrix. By stepwise condensation of the support bound nucleoside with the deoxynucleoside-phosphoramidite monomers elongation of the oligonucleotide chain occurs. The condensation step is followed by a series of further reaction steps i.e. blocking, oxidation and detritylation which are performed heterogeneously. Washing and filtration procedures are necessary after each reaction step in order to remove the mobile reactants. This series of consecutive reaction and washing/filtration steps termed "reaction cycle" has to be carried out for the addition of each single nucleotide unit to the support.

Finally, when the desired sequence is completed, the protecting groups are removed and the oligonucleotide is cleaved from the carrier. The end product is purified by high performance liquid chromatography (hplc) or by polyacrylamide gel electrophoresis (page).

2 EXPERIMENTAL SECTION

2.1 Materials

Chloro-N,N-dimethylamino-methoxyphosphine (prepared as described by Seliger et al., this volume)

1 mmol DMTrdN (N = bzA, bzC, ibG, T)

DMTrdN derivatised silica gel (N = bzA, bzC, ibG, T)

5 L acetonitrile, distilled from CaH_2

2 L nitromethane, stored over 4 Å molecular sieves

5 L tetrahydrofuran containing 0.01 water, stored over 4 Å molecular sieves

200 mL 2,6-lutidine purum, distilled from CaH_2

2 L methanol; 5 L hexane; 2 L chloroform; 1 L n-butanol; 2 L ethylacetate; 1 L toluene; 1 L dioxane; 1 L ammonia (25%); 1 L triethylamine; 1 L acetic acid (96%); 50 mL diisopropylethylamine; 100 mL thiophenol; 100 mL acetic anhydride; 250 g $ZnBr_2$, anhydrous; 100 g iodine suprapur; 100 g N,N-dimethylaminopyridine; 20 g 1H-tetrazole, sublimated at 120°C/0.01 mmHg; 100 g CaH_2; 1 kg NaCl; 1 kg molecular sieve 4 Å; 1 kg basic alumina, 0.1 M triethylammonium acetate (TEAA).

2.2 Equipment

Rotary evaporator; adjustable distilling apparatus with electric heating mantel for a 1 L flask; desiccator; nitrogen cylinder with gauge; hplc with analytical C-18 column; 2 glass columns with sintered glass disc; G3 and G4 glass filter funnels (Schott, Mainz); round-bottomed flasks with ground joint (10 mL, 100 mL, 500 mL); centrifugal glass tubes with lid (15 mL); 5 glass filter funnels Pyrex No. 36060 fine (Fisher Scientific); 5 filter flasks 250 mL; water jet pump; 10 glass syringes 2 mL with long needles; separating funnel 250 mL; mechanical stirrer; Dewars 3-5 L; silicon septa for small flasks or test tubes (see Gait); dry ice/acetone.

3 PROCEDURE

3.1 Synthesis of deoxynucleoside phosphoramidites

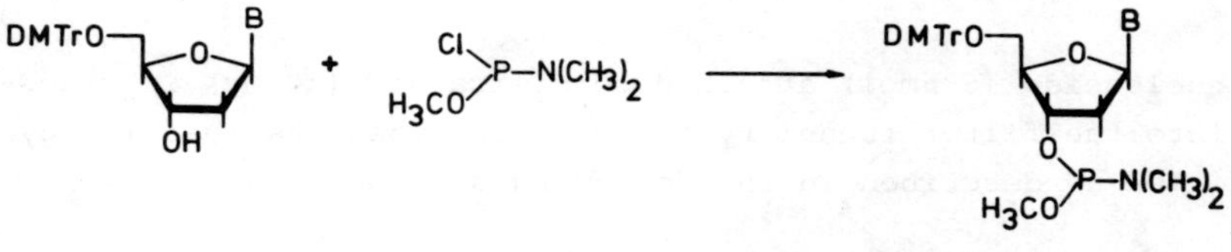

DMTrO–(sugar, B, 3'-OH) + Cl(H$_3$CO)P–N(CH$_3$)$_2$ ⟶ DMTrO–(sugar, B, 3'-O–P(OCH$_3$)–N(CH$_3$)$_2$)

1: B = N-6-benzoyladenine MW: 658.8
2: B = N-4-benzoylcytosine MW: 633.8
3: B = N-2-isobutyrylguanine MW: 639.8
4: B = thymine MW: 544.8
DMTr: 4,4'-dimethoxytrityl

Figure 1 Scheme for the phosphitylation of protected nucleosides

The protected nucleoside (1 mmol 1-4) is dissolved in 3 mL anhydrous, acid-free chloroform and 0.7 mL (4 mmol) diisopropylethylamine in a 10 mL reaction vessel, which is closed with a silicon-septum. The solution is stirred with a magnetic stirrer. 0.2 mL (1.5 mmol) chloro-N,N-dimethylamino-methoxyphosphine (d^{25} = 1.125) is added dropwise by syringe. The solution turns yellow. After 15 min stirring at room temperature the solution is transferred together with 35 mL ethylacetate, to a separating funnel and extracted vigorously four times with 80 mL of a saturated NaCl solution. The pH-value of the

NaCl solution has to be controlled; it has to be above pH 7. The organic phase is dried over sodium sulfate, filtered, and evaporated under reduced pressure. The solid residue is dissolved in 10 mL toluene and added dropwise with vigorous stirring to 50 mL hexane at -78°C. The white precipitate is filtered off through a precooled (-78°C) G3 glass filter funnel and immediately washed with 100 mL cold (-78°C) hexane.

The white powder is dried in a desiccator under vacuum and stored under nitrogen at 4°C. The yield with all four compounds is ∿90%. Products are characterised by ^{31}P-NMR spectroscopy using H_3PO_4 as an internal standard.

		δ-^{31}P (ppm)/$CDCl_3$
DMTr-dbzA-$P(OCH_3)N(CH_3)_2$	:	147.4
DMTr-dbzC-$P(OCH_3)N(CH_3)_2$	:	148.1
DMTr-dibuG-$P(OCH_3)N(CH_3)_2$	:	147.1
DMTr-dT-$P(OCH_3)N(CH_3)_2$	:	147.0

3.2 Reaction cycle of the phosphoramidite method

The nucleoside (5 µmol) attached to silica gel (50 mg) is transferred into the filter funnel Pyrex No 36060. Then the reaction cycle is performed as described in the following section (fig. 2).

step	repetitions	volume [mL]	reagents	time [min]
1	2	1.5	nitromethane	2
2	2	1.5	sat. sol. of $ZnBr_2$ in nitromethane/1% H_2O	5
3	2	1.5	n-butanol:2.6 lutidine:THF (4:1:5, v/v)	5
4	10	1.5	acetonitrile	2
			- glass filter funnel is closed with septum, nitrogen atmosphere	
5	1	0.5	0.5 M tetrazole in acetonitrile	5
		0.5	nucleoside phosphoramidite (100 µmol, 20fold excess) in acetonitrile	5
6	1	1.5	acetonitrile	
7	2	1.5	capping-reagent[a)] 5 mmol dimethylaminopyridine, 10 mmol acetic anhydride, 10 mmol 2.6-lutidine in tetrahydrofuran	5

8	3	1.5	tetrahydrofuran	
			- glass filter funnel is opened	
9	2	1.5	oxidation-reagent[b)]	
		0.2	I_2 in tetrahydrofuran:2.6-lutidine:water (2:1:1. v/v)	5
10	3	1.5	tetrahydrofuran	
	2	1.5	methanol	
	1	1.5	tetrahydrofuran	

a) The capping reagent turns brown in the course of a day. It is prepared freshly every day.

b) The oxidation reagent is filtered before use.

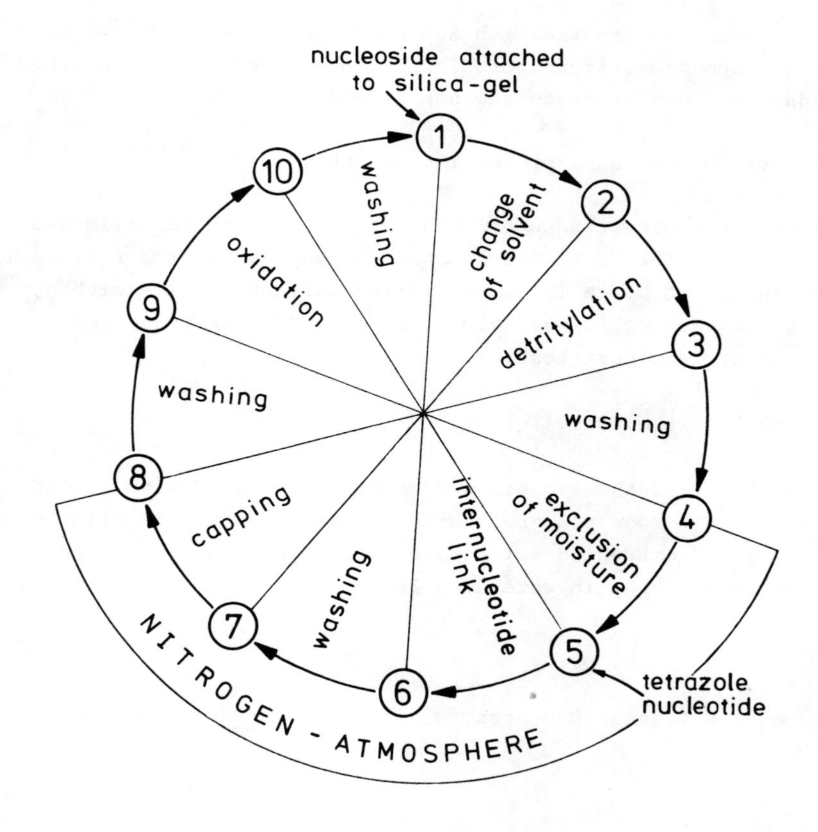

Figure 2 Cycle of the phosphoramidite method

3.3 Removal of protecting groups

3.3.1 Removal of the methyl groups from phosphotriesters

The dry silica gel is transferred from the funnel to a plastic Eppendorf tube. It is treated with 0.5 mL thiophenol:TEA:dioxane (1:1:2, v/v) for 45 min at room temperature. After isolation by centrifugation the silica gel is washed with methanol (4 x 0.5 mL) and ether (2 x 0.5 mL).

3.3.2 Hydrolysis of the ester, removal of the deoxynucleotide from the support

The silica gel is treated with 500 µL concentrated ammonium hydroxide (25%) at room temperature for 3 h. Following centrifugation the supernatant is transferred to another Eppendorf tube.

3.3.3 Removal of the base protecting groups

The tube containing the supernatant is sealed with parafilm and incubated at 50°C for 12 h. The ammonia is evaporated, the residue is dissolved in 0.5 mL 0.1 M TEAA, pH 7.0 and extracted twice with 0.5 mL ether. The oligonucleotide dissolved in 0.1 M TEAA is ready for RP C-18 hplc or page purification.

3.3.4 Removal of the trityl group

For detritylation the lyophilised sample is treated with 0.5 mL 80% acetic acid at room temperature for 15 min. Then it is diluted with H_2O to 2 mL, extracted twice with 5 mL ether and evaporated. The residue is taken up in water and lyophilised.

REFERENCE

[1] Beaucage, S.L. and Caruthers, M.H. (1981) <u>Tetrahedron Lett</u>. <u>22</u>, 1859

PREPARATIVE ISOLATION OF OLIGONUCLEOTIDES FROM CHEMICALLY DEGRADED DNA

Herbert Schott

Institut für Organische Chemie
Universität Tübingen

SUMMARY

A simple alternative to complicated chemical syntheses of oligonucleotides is presented. Readily available DNA is partially hydrolysed to preparative amounts of oligonucleotides using various methods. Pure defined pyrimidine and purine nucleotides, and/or mixtures of sequence isomers, are isolated on a preparative scale from the various mixtures of oligonucleotides by chromatography.

1 INTRODUCTION

As an alternative to the complicated chemical methods available, we have begun to develop a simple strategy for preparing oligonucleotides without synthesis. Although this method is limited to oligonucleotides of given natural sequences, it is much easier to perform than chemical synthesis. According to this concept DNA is partially hydrolysed by various chemical methods designed to destroy selectively certain monomer units. The chemical degradation is based on analytical procedures [1-6], which we have adapted to preparative methods. The cleavage reactions currently possible are summarised in scheme 1.

--- A-T-T-C-G-T-G-A-A-A-C-C-T-T-T-G-C-G-G-T-T-A-A-A-A ---

	Methods		Hydrolysates
a)	HCOOH + diphenylamine		T-T-C, T, C-C-T-T-T, C, T-T
b)	1. $KMnO_4$	2. OH^-	A, A-A-A, A-A-A-A
c)	1. NH_2NH_2	2. OH^-	A, G, G-A-A-A, G-G, A-A-A-A
d)	1. NH_2NH_2	2. OH^-	
	3. piperidine	4. OH^-	G, G-G
e)	pyrimidine nucleotides from a)		
	1. OH^-	2. NH_2NH_2	T-T, T, T-T-T
f)	purine nucleotides from c)		
	1. OH^-	2. piperidine	
	3. OH^-		G, G-G

Scheme 1 Chemical partial hydrolysis of DNA

The chemical partial hydrolysis yields highly complex mixtures which can be almost completely separated by chromatography. Known chromatographic methods (ion exchange-, template-, paper-, and hplc) are used to separate the partial hydrolysates into mixtures of determinable composition, and/or DNA fragments of defined sequences. At present over 200 defined sequences with 2-12 monomer units can be obtained from the various partial hydrolysates of DNA by a few separation steps [7-19].

2 EXPERIMENTAL SECTION

2.1 Materials

350 mg preseparated pyrimidine oligonucleotides of peak IVa, fig.2
150 mL QAE-Sephadex A-25 (column: 2 cm x 48 cm)
equilibrated with 200 mM NaCl, 50 mM Tris-HCl, pH 7.5
7 M urea
3 L 200 mM NaCl, 50 mM Tris-HCl, pH 7.5, 7 M urea
1 L 1 M NaCl
20 mL 7 M urea
20 mL ca. 1% $AgNO_3$ solution in water
400 mL Sephadex G-10 (column 3 cm x 60 cm) in water
300 mL System A: n-propanol:NH_3 conc.:water (55:10:35, v/v)
300 mL System B: C_2H_5OH:1 M ammonium acetate, pH 7.5 (7:3, v/v)
10 mL 1 M Tris-HCl, pH 8-10
10 µL alkaline phosphatase (2000 U/mL) from calf thymus
chromatography paper: Schleicher & Schüll Nr. 2316
(58 cm x 60 cm)

2.2 Equipment

a) equipment for column chromatography (pump)
b) rotary evaporator
c) uv-spectrometer
d) lyophyliser
e) solvent tank for descending paper chromatography (58 cm x 60 cm)
f) water bath 37°C

Total time: 48 h
Working time: 8 h

2.3 Procedure

2.3.1 Isolation of $p(dT)_3p$

In the following example the isolation of $p(dT)_3p$ is performed.

2.3.1.1 By treatment of herring sperm with formic acid in the presence of diphenylamine (scheme 1a) the purine monomer units are destroyed. The pyrimidine nucleotide segments, however, remain intact and are released from the depurinated DNA [7,8].

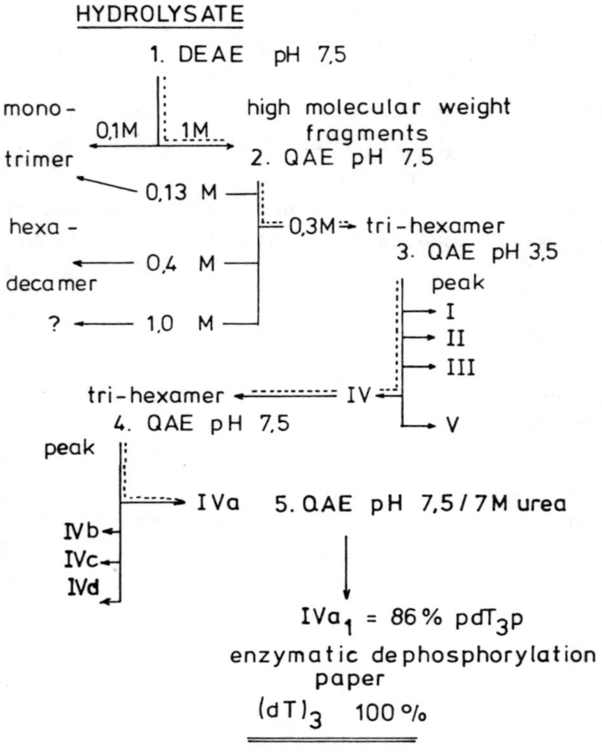

HYDROLYSATE

1. DEAE pH 7,5

mono- trimer ← 0,1M | 1M → high molecular weight fragments

2. QAE pH 7,5

trimer ← 0,13 M

hexa - decamer ← 0,4 M

? ← 1,0 M

0,3M → tri-hexamer

3. QAE pH 3,5

peak

I

II

III

IV

V

tri-hexamer ← IV

4. QAE pH 7,5

peak

IVa

IVb

IVc

IVd

5. QAE pH 7,5 / 7M urea

IVa_1 = 86 % pdT_3p

enzymatic dephosphorylation

paper

$(dT)_3$ 100 %

Scheme 2 Route of separation used for isolating monopyrimidine to decapyrimidine nucleotides from the partial hydrolysate of depurinated DNA

2.3.1.2 From the resulting partial hydrolysates defined pyrimidine oligonucleotides (up to decamers) are isolated chromatographically according to scheme 2 [7,8,10,11,12,14,16].

Preliminary separation of the partial hydrolysate is performed on a DEAE-cellulose column employing a two-step gradient of NaCl buffered by Tris-HCl to pH 7.5. In the first gradient step (0.1 M NaCl) most of the low molecular weight oligonucleotides (up to trimers) are eluted. In the second gradient step the small amount of higher molecular fragments are eluted with 1 M NaCl. The fractions are desalted by ultrafiltration. In a subsequent separation the fractions (0.1 M and 1 M) are rechromatographed on a QAE-Sephadex column. Pure monomer to trimer pyrimidine nucleotides can be isolated from the 0.1 M fraction [7,8]. The 1 M fraction yields up to decameric pyrimidine oligonucleotides according to the following separation route.

The 1 M fraction is separated into 4 fractions using a four-step gradient of NaCl (0.13 M, 0.3 M, 0.4 M, 1 M) at pH 7.5. The 0.13 M fraction contains the rest of the 0.1 M fraction (up to trimers with one terminal phosphate group). The products of the 0.3 fraction are trimers (with two terminal phosphate groups) to hexamers [12]. The 0.4 M fraction contains a mixture of hexameric to decameric pyrimidine oligonucleotides. The 1 M fraction has not yet been investigated.

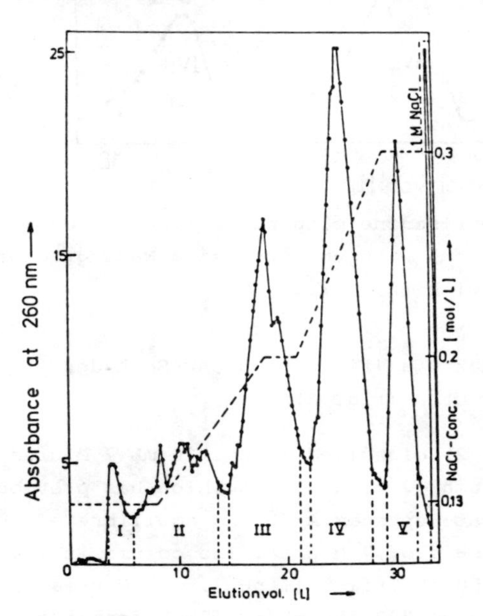

Figure 1 Separation of 15 g pyrimidine oligonucleotides on QAE-Sephadex (60 cm x 5 cm) with an increasing NaCl gradient, buffered at pH 3.5 with 50 mM NaOAc

The desalted and lyophylised 0.3 M fraction is then separated on a QAE-Sephadex column with a six-step gradient of NaCl at pH 3.5 into 5 peaks (I-V) according to the type of thymine bases (fig. 1). Peak IV contains pyrimidine oligonucleotides with identical thymidylate but with varying cytidylate units (dC_{0-3}, dT_3). In step 4 they are separated on QAE-Sephadex at pH 7.5 into 4 peaks (IVa-d) according to the number of phosphate charges (fig. 2). The following procedure is carried out in order to isolate $p(dT)_3p$ from peak IVa shown in fig. 2.

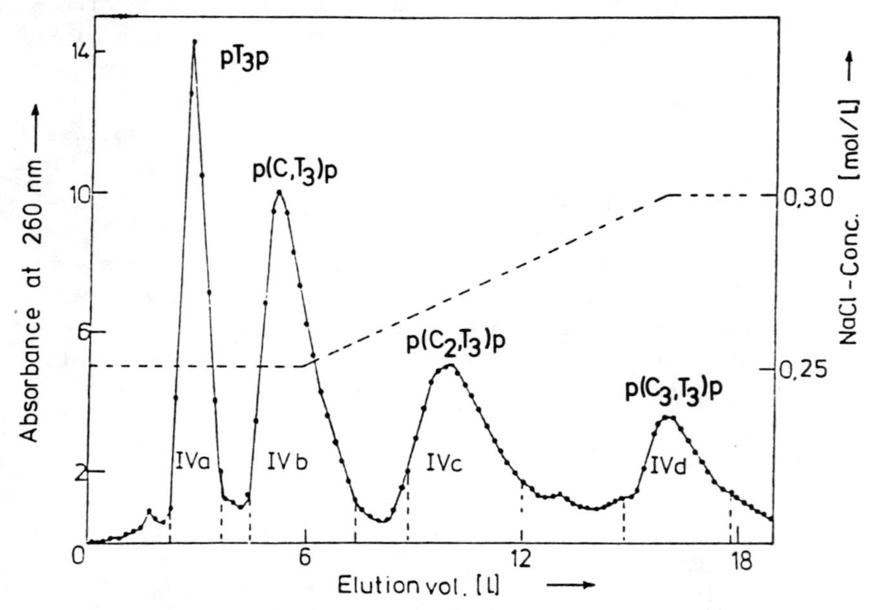

Figure 2 Separation of 3 g pyrimidine oligonucleotides on QAE-Sephadex (70 cm x 3 cm) with an increasing NaCl gradient, buffered at pH 7.5 with 50 mM Tris-HCl

2.3.2 Rechromatography of peak IVa (fig. 2) on QAE-Sephadex at pH 7.5 in the presence of 7 M urea (fig. 3)

350 mg from peak IVa (fig. 2) dissolved in ca. 10 mL 7 M urea are applied to the top of the column (2 cm x 48 cm) which has previously been filled with ca. 130 mL QAE-Sephadex A-25 and equilibrated in 200 mM NaCl, 50 mM Tris-HCl, pH 7.5, 7 M urea. The column is eluted with ca. 2 L of 200 mM NaCl, 50 mM Tris-HCl, pH 7.5, 7 M urea. The flow rate of the column should be 200-300 mL/h. 15 mL fractions are collected. The eluate is monitored by an automatic uv-recorder and by measurement of the absorbancy of individual fractions in a spectro-

photometer. The fractions of IVa_1 (fig. 3) containing the desired $p(dT)_3p$ are combined between the dotted lines and desalted using the following procedure.

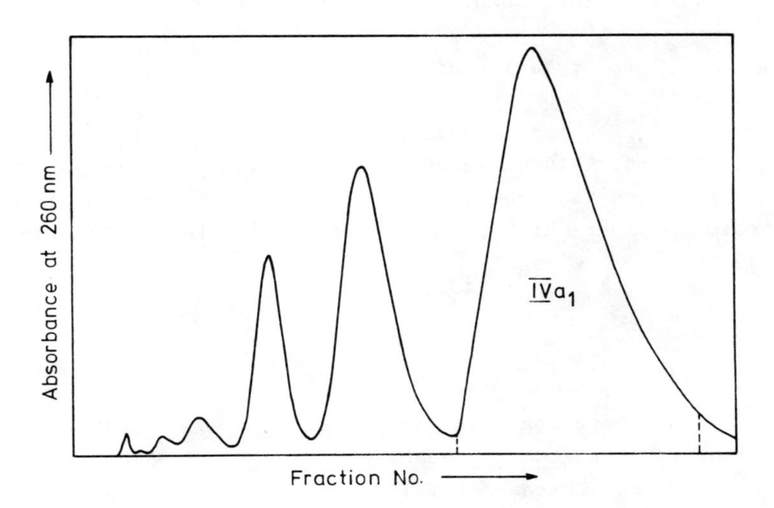

Figure 3 Separation of 350 mg of peak IVa (fig. 2) on QAE-Sephadex (2 cm x 48 cm) with 0.2 M NaCl, 0.05 M Tris-HCl, pH 7.5, 7 M urea

2.4 Desalting of peak IVa_1 (fig. 3)

The column from 2.3.2 is eluted with water until NaCl can no longer be detected in the eluent by $AgNO_3$. The combined fractions (∿ 500 mL) of peak IVa_1 (fig. 3) are diluted with water to a 2.5 fold volume and again applied to the QAE-Sephadex A-25 column from 2.3.2. The column is washed free of urea with water. In a subsequent step adsorbed $p(dT)_3p$ is eluted with 1 M NaCl. The fractions of $p(dT)_3p$ are concentrated with a rotary evaporator until NaCl precipitates. The precipitate is dissolved by adding a small amount of water, and $p(dT)_3p$ is desalted by passing it over a Sephadex G-10 column (3 cm x 60 cm, flow rate 80 mL/h). The oligonucleotide is eluted in the void volume. The volume applied for desalting should not exceed ca. 10 mL. The oligonucleotide is concentrated with a rotary evaporator and lyophilised. 200 mg of $p(dT)_3p$ are obtained as a white powder of 98% purity.

Characterisation of $p(dT)_3p$:

a) uv: $A_{250}/A_{260} = 0.66$; $A_{280}/A_{260} = 0.69$

b) R_F relative to pdT in system A = 0.50
System A: n-propanol:NH_3 conc.:H_2O (55:10:35, v/v)

2.5 Enzymatic dephosphorylation of $p(dT)_3p$

1-2 mg (ca. 30 A_{260} units) are dissolved in ca. 50 µL H_2O and 10 µL 1 M Tris-HCl, pH 8-10. 5-10 µL (2000 U/mL) alkaline phosphatase are added and the solution is incubated at 37°C for 2-4 h. The solution is chromatographed on paper which is developed with system B. R_F relative to pdT = 1.42.

System B: C_2H_5OH:1 M ammonium acetate, pH 7.5 (7:3, v/v).

REFERENCES

[1] Burton, K. and Peterson, G.B. (1960) Biochem. J. 75, 17
[2] Habermann, V. (1963) Coll. Czech. Chem. Comm. 28, 510
[3] Shapiro, H.S. and Chargaff, E. (1964) Biochim. Biophys. Acta 91, 262
[4] Burton, K. (1967) Methods Enzymol. 12A, 222
[5] Brammer, K.W., Jones, A.S., Mian, A.M., and Walker, R.T. (1968) Biochim. Biophys. Acta 166, 732
[6] Maxam, A. and Gilbert, W. (1977) Proc. Natl. Acad. Sci. USA 157, 197
[7] Schott, H. and Schwarz, M. (1977) Hoppe-Seyler's Z. Physiol. Chem. 358, 949
[8] Schott, H. and Schwarz, M. (1978) Hoppe-Seyler's Z. Physiol. Chem. 359, 617
[9] Schott, H. and Schwarz, M. (1978) J. Chromatogr. 157, 197
[10] Schott, H. (1978) Nucl. Acids Res. Spec. Publ. 4, 161
[11] Dizdaroglu, M., Hermes, W., von Sonntag, C., and Schott, H. (1979) J. Chromatogr. 169, 429
[12] Schott, H. (1979) J. Chromatogr. 172, 179
[13] Schott, H. (1980) J. Chromatogr. 187, 119
[14] Dizdaroglu, M., Simic, M.G., and Schott, H. (1980) J. Chromatogr. 188, 273
[15] Schott, H. and Watzlawick, H. (1980) J. Chromatogr. 196, 435
[16] Schott, H. (1980) Nucl. Acids Res., Symposium Series 7, 203
[17] Schott, H. (1981) Makrom. Chem. 182, 2015
[18] Schott, H. and Schrade, H. (1981) Nucl. Acids Res., Symposium Series 9, 187
[19] Schott, H. (1982) J. Chromatogr. 237, 429

PRINCIPLES OF AUTOMATED GENE FRAGMENT SYNTHESIS

Ronald M. Cook, Derck Hudson, Eric Mayran, and Jonathan Ott

Biosearch
San Rafael

SUMMARY

Automated chemical gene fragment synthesis has its roots in the solid-phase peptide methodology as pioneered by R.B. Merrifield. Meanwhile the development of strategies for the synthesis of internucleotide bonds has led to effective solid phase procedures eventually allowing automation. The basic components of an automated oligonucleotide synthesizer in terms of quality parameters are outlined, and requirements for chemistry, solvents and mechanical and electronic parts are discussed. The major advantage of automated synthesizers over manual or semi-automated techniques are the reduced costs for labor. Further developments in the field will concentrate on automated RNA fragment synthesis, longer oligonucleotides, purer fragments, integrated synthesis and purification and finally an analytical feedback system.

1 INTRODUCTION

The feasibility of automated gene fragment synthesis is the result of significant developments in the fields of peptide chemistry, nucleotide chemistry and semiconductor design and manufacture. The introduction of solid-phase peptide synthesis by R.B. Merrifield [1], R.L. Letsinger and M.J. Kornet [2] demonstrated the possibility of simplifying organic synthesis to a sufficient degree to allow eventual automation. Chemical synthesis of internucleotide bonds has in the past two decades been refined to effective solid-phase procedures. Meanwhile, advances in semiconductor design and manufacture have led to mass production of inexpensive microprocessors capable of reliable and efficient control of solid-phase chemical reaction systems.

The development of strategies for synthesizing internucleotide bonds has been well covered by other communications in this volume, and is beyond the scope of this review. Only solid-phase methodologies are readily amenable to automation. The most popular solid-phase oligonucleotide synthesis methods are the phosphate-triester approach, depicted in figure 1, and the phosphite-triester method, shown in figure 2.

2 HISTORY OF AUTOMATED CHEMISTRY

After establishing the principle of solid-phase peptide synthesis R.B. Merrifield initiated work on an automated synthesis system. J.M. Stewart, in collaboration with Merrifield, constructed a machine with a shaken reactor vessel which contained the solid support, and a metering pump to dispense reagents selected via rotary teflon valves from a bank of reagent and solvent reservoirs. The system was controlled by a timed rotating metal drum (much like that on a music box) triggering up to 30 microswitches according to the placement of nylon pins on the drum's surface. The first working model was developed in 1965 (see figure 3), and featured a cycle time of 4 h per amino acid residue [3,4]. By 1969, R.B. Merrifield and B. Gutte had used the machine in the synthesis of ribonuclease A, involving 369 reactions representing 11,391 steps by the machine [5]. In the past decade, a number of commercial peptide synthesizers employing the basic principles adumbrated by Stewart and Merrifield have been developed and marketed.

Automated synthesis of oligonucleotides was not attempted until the late 1970's. Solid-phase nucleotide synthesis is fundamentally similar to solid-phase peptide synthesis in that it involves repetitive addition of activated, blocked residues to an insoluble polymer support.

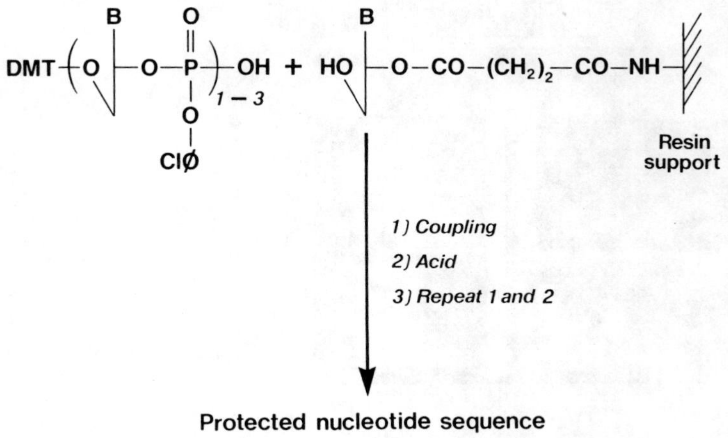

$$DMT\left(O\ \text{[B]}\ O-\overset{O}{\overset{\|}{P}}(OC_6H_4Cl)\right)_{1-3}-OH + HO\ \text{[B]}\ O-CO-(CH_2)_2-CO-NH-\text{Resin support}$$

1) Coupling
2) Acid
3) Repeat 1 and 2

Protected nucleotide sequence

Figure 1 Phosphate-triester method
Cycle time 1.5 h, coupling time 1 h

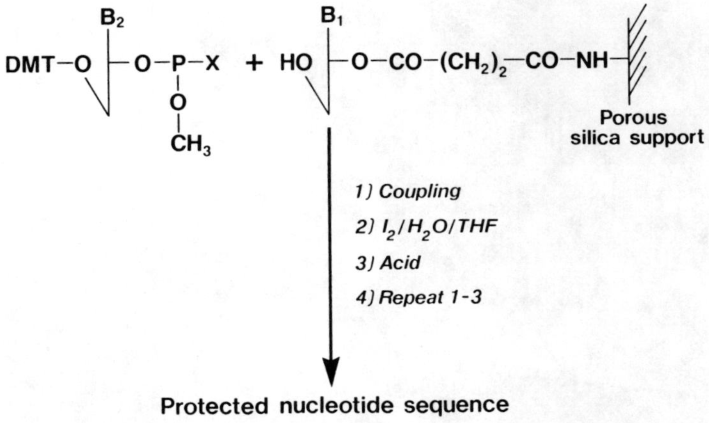

$$DMT-O\ \text{[}B_2\text{]}\ O-P(OCH_3)-X + HO\ \text{[}B_1\text{]}\ O-CO-(CH_2)_2-CO-NH-\text{Porous silica support}$$

1) Coupling
2) $I_2/H_2O/THF$
3) Acid
4) Repeat 1-3

Protected nucleotide sequence

Figure 2 Phosphite-triester method
Cycle time 30 min, coupling time 10 min. X = -Cl, -tetrazole, or $-N(CH_3)_2$

Figure 3 J.M. Stewart's original automated peptide synthesizer. Note "music box" controller. Photograph from Solid Phase Peptide Synthesis by Stewart and J.D. Young (Freeman, San Francisco, 1969)

Figure 4 Author's manual gene fragment synthesis apparatus. Note pump and hplc guard column. Test tubes contain detritylation effluent

However, automation of the former had to await refinenment of solid-phase nucleotide coupling methods. High coupling efficiencies (at least 90% per coupling) are desirable for successful synthesis of 15-20 base fragments. By 1980, numerous procedures were available for automation, and to date a handful of firms have marketed or announced automated oligonucleotide synthesizer systems. Such systems have employed the phosphate-triester method, as well as the phosphite-triester method, using chloridite, amidite, and tetrazolide intermediates. The phosphate-triester method involves generally lower coupling efficiencies than the phosphite-triester approach. Use of MSNT as a coupling agent provides yields high enough to couple blocked monomers by this approach. In the case of lower-yielding reactions, blocked dimers or trimers may be coupled, decreasing the number of coupling steps. This technique has the added advantage of increasing the feasible length of the oligonucleotide product. However, when monomers are coupled, the machine needs to select from only four base reservoirs, whereas if dimers or trimers are employed, the number of reservoirs sampled would be considerably higher for some syntheses.

The most important innovation employed in commercial oligonucleotide synthesizers has been microprocessor control. It is unfortunate that the earliest microprocessor-controlled gene fragment synthesizers were marketed before the hardware and chemistry were perfected, giving automated gene fragment synthesis an undeserved bad reputation.

3 MANUAL, SEMI-AUTOMATIC AND AUTOMATIC GENE FRAGMENT SYNTHESIS

Techniques for gene fragment synthesis can be categorized as manual, semi-automatic, or automatic. Regardless of the technique chosen, success of the synthesis depends on using pure starting materials and dry solvents. M. Gait (see this volume) has suggested the concept of "impurity amplification" - should a nucleotide agent have 1% of impurity and be used in the reaction in ten-fold excess, the impurity may become concentrated in the reaction, leading to drastically lowered yields. Some suppliers of starting materials for gene fragment synthesis have chosen to offer pre-packaged reagent kits. This strategy has not yet proven successful, and knowledgeable scientists might best prepare their own starting materials, regardless of the synthetic method chosen.

Manual synthesis is defined as any synthetic method in which the scientist must handle the mixing, decanting, filtering and other manipulations of the solvents and reagents. This method can employ apparatus a simple as a test tube or sintered glass funnel, but may also employ a pump and column, as depicted in figure 4. The key distinction here is that constant attention on the operator's part is required, and the operator must himself know in detail the steps of the synthesis.

Semi-automatic synthesis employs an automatic solvent delivery device such as an analytical hplc unit with binary (or prefereably ternary) solvent handling capabilities. The authors have employed a Varian 5000 microprocessor-controlled hplc in the synthesis of oligonucleotides. In this case, constant attention by the operator is not required, but the nucleotides must be injected via the sample loop at each base step, so that each cycle requires one manual intervention by the operator. Again, the operator must know fundamentally the technique of the synthesis. A system like the Varian 5000 is just as good a nucleotide synthesizer as some of the dedicated gene fragment synthesizers on the market, which also require periodic injection of nucleotides. The microprocessor modulates the pump flow to maximum effect for each step, according to operator programming. In theory such a system could be made fully automatic by the addition of an automatic sample injector device.

Automatic gene fragment synthesizers have the capability of automatic injection of the nucleotides as well as automatic sampling of the necessary solvents and reagents. In this case, the machine is preprogrammed for all aspects of the synthesis, so the operator need not know the synthetic method. Moreover, the operator need not be present at all during the synthesis. He need only fill the reagent and solvent reservoirs and start the program. A "home-made" version of an automatic gene fragment synthesizer can be assembled using virtually any microcomputer as controller, with an output interface card to switch valves and control the pump. Figure 5 shows a system built by the authors around an Apple II computer. Such a system could be assembled for approximately $ 5000 ($ 2500 for the computer with monitor, disk drive, and printer, $ 400 for the interface card, $ 1500 for valves, and $ 800 for the pump). Even this basic microcomputer has computer power greatly in excess of that needed for controlling a gene fragment synthesis apparatus. The authors will gladly supply on request a complete program suitable for oligonucleotide synthesis control by an Apple II computer.

4 PRINCIPLES OF AUTOMATED GENE FRAGMENT SYNTHESIS

Figure 6 illustrates schematically an automated gene fragment synthesizer, which has the following basic components:

Figure 5 Author's fully automatic gene fragment synthesis apparatus employing an Apple II microcomputer as controller. Dot matrix printer not shown. Such a system embodies all the desirable characteristics of an automated synthesizer

1. controller;
2. reagent and solvent delivery system;
3. servo-mechanical valves;
4. solid-phase reactor.

An automated gene fragment synthesizer is basically a precise, timed, solvent dispensing apparatus. The chemical capabilities of the machine are a function of the programming and the precursors, solvents, and reagents used. Refinements and improvements in automated gene fragment synthesis must follow chiefly from chemical research and development. The optimum parameters for automated gene fragment synthesis are as follows:

1. Controller - While in principle a mechanical "music box" controller (such as used in Stewart's original peptide machine) will serve the purpose, optimum performance demands a microprocessor controller. Having no moving parts, the microprocessor has the not inconsiderable advantage of greater reliability than its predecessor. Moreover, a microprocessor incorporating approximately 4 K bytes of memory (RAM or ROM) has the capability of simultaneous storage of numerous programs, programming the microprocessor requires only a few keystrokes, and programs can easily be modified during execution. Finally, a microprocessor controller is less expensive than its mechanical predecessor. Table 1 shows a program for phosphite-mediated nucleotide synthesis using a commercially-available gene fragment synthesizer.
2. Reagent and Solvent Delivery System - Pressurized inert gases (such as nitrogen and argon) and teflon or glass tubing are adequate for reagent and solvent delivery. Some systems, however, also employ a pump. The pump is placed downstream of the valves, while pressurized inert gas is used to dispense solvents and reagents to the valves and ultimately to the pump. The pump provides the precise flow rates consistent with high performance. While conventional pumps will give good service, optimum performance demands a stepper motor pump. Prerequisites for solvent delivery systems are lack of cross-contamination between reagent vessels, and minimal dead volume in all plumbing.
3. Servo-Mechanical Valves - Some automated synthesizer machines have used pneumatically-actuated valves. Optimum performance requires electric solenoid-actuated valves. In either case. solvent-accessible surfaces should be inert glass or teflon. Ideally, the valves would have zero dead volume. In any system, great care must be taken to prevent particulate matter from clogging the valves; use of in-line filters is necessary, and reagents and solvents must be selected so as to prevent in-line crystallization.

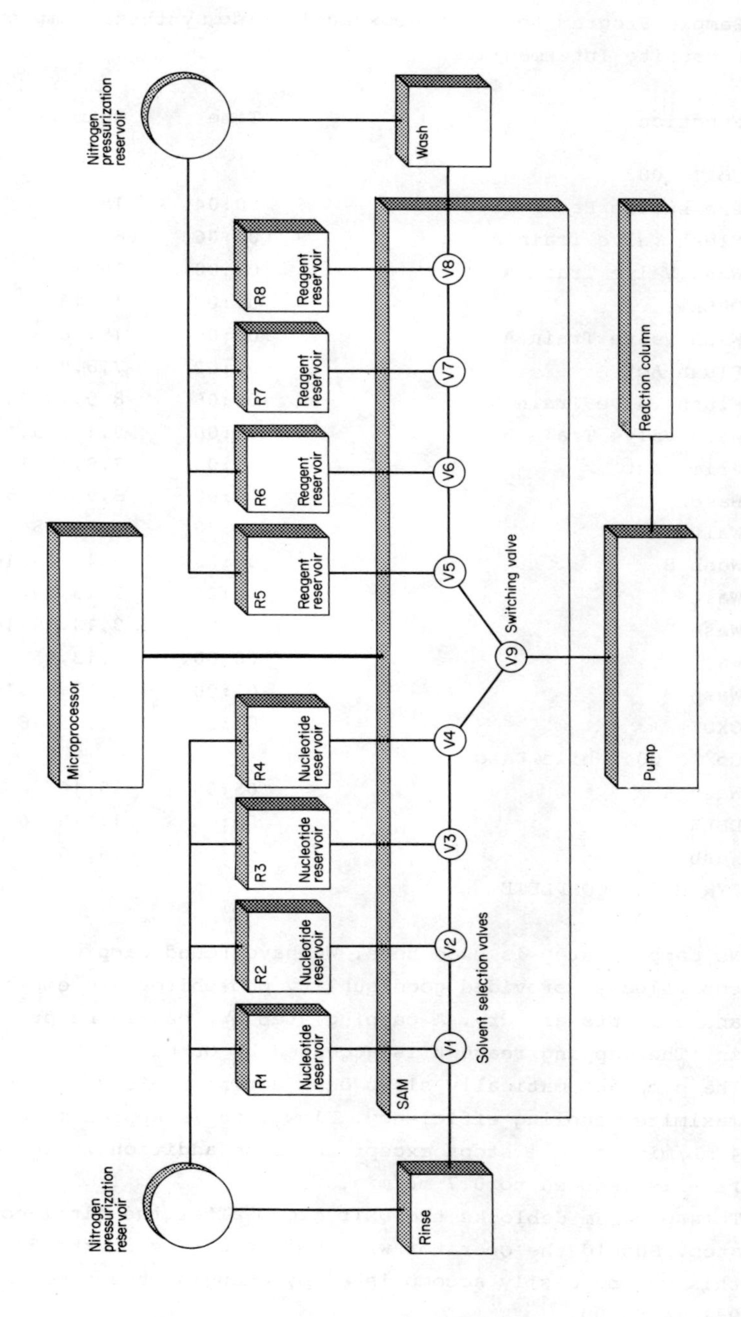

Figure 6 Schematic diagram of microprocessor-controlled, automated gene fragment synthesizer

Table 1 Sample Program for Oligodeoxynucleotide Synthesis Employing Phosphite Intermediates

Step #	Function	Time	Outputs
001	Go To 002		
002	Pressurize Reservoirs	00:04	15
003	Flush Valve Train A	00:06	8,15
004	Wash Valve Train A	05:00	15,16
005	DBLK	02:00	11,15,16
006	Wash Valve Train A	03:00	15,16
007	Flush Act	00:02	7,8,9,15
008	Flush Valve Train B	00:06	8,9,13,15
009	Wash Valve Train B	01:00	9,13,15,16
010	Prime Act	00:05	7,9,15,16
011	Base	04:00	B,9,15,16
012	Wait	00:30	9,13,15
013	Wash B	00:05	9,13,15,16
014	Wait	00:30	9,13,15
015	Wash B	00:05	9,13,15,16
016	Wait	00:30	9,13,15
017	Wash B	03:00	9,13,15,16
018	OXDZ	01:00	10,15,16
019	Go To 004 While Base		
020	Wash A	05:00	15,16
021	DBLK	02:00	11,15,16
022	Wash A	04:00	15,16
023	SYNTHESIS COMPLETE		

NOTE: A. No capping step is used here. We have found capping to be superfluous, provided good quality phosphites are employed and solvents are dry. A capping step may easily be programmed in. The capping reagent is accessed by Output 12.

B. The pump automatically slows down during the coupling step to maximize coupling efficiency. Flow rate is approximately 4 mL/min for all steps except the base addition in which the rate is reduced to 0.7 mL/min.

C. This program deblocks the DMTr group after the final coupling step. Should the operator wish DMTr blocking of the 5' residue, this can be easily accomplished by changing the time for step 021 to 00:00.

4. Solid-Phase Reactor - Both a stirred or shaken, static reactor vessel, and a fritted column continuous-flow reactor have been employed in automated gene fragment synthesis. While the continuous-flow approach is simpler and requires less hardware, some chemical procedures can in principle be accomplished more economically (due to slow reaction rates) with a static reactor. However, with the addition of a few extra valves, a continuous-flow system can be adapted to recycle reagents over the column, thus making economical use of solvents even in slow reaction steps.

5 ADVANTAGES OF AUTOMATED GENE FRAGMENT SYNTHESIS

Automated gene fragment synthesizers can accomplish only what conventional manual methods can achieve - the synthesis of a crude blocked oligonucleotide preparation containing various truncated sequences and other by-products in addition to the desired gene fragment. Contrary to reports in the press, these machines do not synthesize DNA or genes, but rather crude blocked preparations containing DNA or gene fragments, which require manual deblocking and substantial manual purification either by hplc or preparative polyacrylamide gel electrophoresis before they are suitable for use in biological experiments. What, then, are the advantages of automated gene fragment synthesizers?

The main advantage of automated synthesis lies in its capacity to save time, labor and ultimately money. With a nominal amount of set-up time (given an efficiently designed system) a synthesis can be accomplished that otherwise would require 8 h or more of tedious and painstaking manipulations. Further time and labor can be saved by running automated syntheses overnight. The automated system effectively transfers the work load from synthetic manipulations, freeing up time for the more time-consuming task of purification of the product.

Another major advantage of automated synthesis is that a properly designed and programmed synthesizer machine encapsulates the technology and techniques required for gene fragment synthesis. While any technician could be trained in a few hours to run an efficient gene fragment synthesizer, only a capable and diligent technician with some days or weeks of specialized training (which may not be available in many molecular biology laboratories) could accomplish manual gene fragment synthesis. Furthermore, repetitive manual syntheses are tedious and require constant attention. Few chemists or technicians would wish to synthesize gene fragments repeatedly by manual means.

An additional advantage of efficient automated systems is repeatability. A well designed and maintained machine is less prone to err in tedious repetitive procedures than is a human being. Accordingly, automated synthesizers maximize the operator's ability to control variables in experimental procedures.

Finally, an automated system should be able to outperform manual synthesis because it carries out the synthesis in a closed system, effectively isolating the reagents from air and moisture which can lead to decomposition of reagents and diminished yields.

6 DISADVANTAGES OF AUTOMATED GENE FRAGMENT SYNTHESIS

Automated gene fragments synthesizers impose certain restrictions on synthetic parameters that would not pertain to manual synthesis methods. In automated systems, all reagents must be readily soluble so as not to crystallize in the system, leading to clogging of tubing and destruction of valves. The reagents must also be stable for the duration of the synthesis, and the viscosity of the reagents is restricted. Viscous solvents do not pump well, can cause high back pressures on the column, and even clog tubing. Some solvents, like ether, may cause cavitation of the pump. All reagents must be inert to the wetted surface materials of the machine. Some nucleotide reagents, such as chloridites may in time prove corrosive to stainless steel used in pumps. Finally, synthetic methods requiring heterogeneous reaction phases are not easily automated.

7 ECONOMICS OF AUTOMATED GENE FRAGMENT SYNTHESIS

Automated gene fragment synthesizers are currently available commercially, at prices ranging from $ 18,000 to $ 40,000. Since the machine accomplishes nothing that cannot be accomplished by manual means, and because both methods use virtually the same amounts of raw materials and yield products requiring the same purification procedures, the worth of the machine must be assessed as a function of labor saved. An automated system can accomplish one synthesis during a working day, and be set up to run another overnight. Two man-days would be required to accomplish these syntheses manually. At $ 100 (labor and benefits) per man-day, each day the machine is used will result in a $ 200 savings in labor. Thus the machine would pay for itself after 90 to 200 days of use (purchase price $ 18,000 - $ 40,000); in other words, after 180 - 400 syntheses had been run.

8 FUTURE DEVELOPMENTS IN AUTOMATED GENE FRAGMENT SYNTHESIS

A number of anticipated future developments in automated gene fragment synthesis will make the machines even more valuable tools in molecular biology research.

1. RNA Fragment Synthesis - Programs should soon be available to allow automated RNA fragment synthesis. This would employ virtually the same hardware as current DNA fragment synthesis.
2. Capability of Making Longer Fragments - Current methods are adequate to the synthesis of 15 - 20 base gene fragments. Improved supports and coupling efficiencies will likely make synthesis of 30 - 40 base fragments possible.
3. New Chemical Methods - While there are presently several methods employed for automated synthesis of internucleotide bonds, others will be introduced, notably the hydroxy-benzotriazole method of Van Boom.
4. Multiple Synthesis Capabilities - Hardware and software improvements should make it possible to run multiple simultaneous syntheses, further improving the economics of automated synthesis.
5. Integrated Synthesis and Purification - The next 2 - 3 years will likely see the introduction of multi-column machines, in which the product will be cleaved from the reaction column and passed to a deblocking reactor, before being shunted to one or more purification columns, and finally to a detector and microprocessor-controlled fraction collector.
6. Reduced Operational Costs - Currently the cost of preparing a nucleotide 15-mer using commercial synthesizers is between $ 200 and $ 1000, depending on the machine chosen. Competition should drive the average cost down closer to the lower figure.
7. Analytical Feedback - Machines will eventually employ some analytical feedback system, such as colorimetric analysis of the detritylation effluent, to have the microprocessor compute yields of individual coupling steps, and print a record of these data, for use in troubleshooting in the event of problems.

9 REFERENCES

[1] Merrifield, R.B. (1963) J. Am. Chem. Soc. 85, 2149

[2] Letsinger, R.L. and Kornet, M.J. (1963) J. Am. Chem. Soc. 85, 3045

[3] Henahan, J.F. (1971) Chem. Eng. News 2, 22

[4] Stewart, J.M. and Young, J.D. (1969) Solid Phase Peptide Synthesis, Freeman

[5] Merrifield, R.B. and Gutte, B. (1971) J. Biol. Chem. 246, 1922

ADDITION OF HOMOPOLYMER TRACTS TO SINGLE STRANDED AND DOUBLE STRANDED DNA FRAGMENTS BY TERMINAL DEOXYNUCLEOTIDYL TRANSFERASE

Marion Schmitt and Hans Günter Gassen

Institut für Organische Chemie und Biochemie
Technische Hochschule Darmstadt

SUMMARY

Depending on the presence of either Mg^{2+} or Co^{2+} ions the enzyme terminal deoxynucleotidyl transferase may be used to add a number, preferentially 15-20 of nucleotide residues to the 3'-end of either single stranded or double stranded DNA fragments. This article gives the experimental details for the addition of oligocytidylate or oligoguanylate tracts to chemically prepared single stranded oligodeoxynucleotides and to the linearized plasmid pBR322.

1 INTRODUCTION

The enzyme terminal deoxynucleotidyl transferase, usually called TdT or Bollum's enzyme [1], catalalyses the polymerisation of deoxynucleotide residues at the 3'-termini of single stranded DNA or oligodeoxynucleotides [2]. For catalytic activity the enzyme requires a free 3'-OH group and a minimum of three nucleotide residues in a single stranded configuration. One unit is defined as the amount of protein which incorporates 1 nmol dATP per hour at 37°C into acid precipitable form using $p(dA)_{\sim 50}$ as primer [3]. The enzyme also catalyses a limited polymerisation of ribonucleotides at the 3'-end of oligodeoxynucleotides [4]. Such ribonucleotide addition has been shown to be useful for primer extension [5-7], 3'-end labeling [8,9] and DNA sequence analysis [10-13]. Furthermore TdT can be used to radiolabel the 3'-termini of oligodeoxynucleotides with ^{32}P-cordycepin-5'-triphosphate (3'dATP) for DNA sequence determination [3]. Recently TdT has proved useful to tail single stranded copy DNA (cDNA) obtained from mRNA by reverse transcription [14]. cDNA tailed with a tract of cytidylates may then be double stranded with an oligo-G primer and reverse transcriptase. TpT can accept double stranded DNA as primer for the addition of deoxynucleotides and ribonucleotides if Mg^{2+} is replaced by Co^{2+}. All forms of duplex DNA molecules can be labeled at their unique 3'-end regardless of whether such ends are staggered or even [15]. Therefore without prior treatment with exonuclease to expose the 3'-terminus as single stranded primer, this reaction now permits addition of homopolymer tails at the 3'-ends of all types of DNA molecules for the purpose of *in vitro* construction of recombinant DNA. The kinetics of the reaction can be controlled such that on the average 15-20 nucleotides are added to the DNA molecule, which represents the optimal protruding end for the consecutive ligation reaction.

2 EXPERIMENTAL SECTION

2.1 Reagents and equipment

Restriction endonuclease *Bam* H1 (EC 3.1.23.6); terminal deoxynucleotidyl transferase (from BRL); pBR322; synthetic oligonucleotide n = 12-20; dCTP; [α-^{32}P]dCTP; dGTP; [α-^{32}P]dGTP; cacodylate-K; $MgCl_2$; dithiothreitol (DTT); EDTA-Na_2; NaOAc; Tris-HCl; $CoCl_2$; NaCl; sodium dodecylsulfate (SDS); trichloroacetic acid (TCA); chloroform; isoamylacetate; phenol; Eppendorf vials; glass fiber filters Whatman GF/A; 20-200 μL automatic pipettes; 1-5 μL glass micropipettes; 37°C and 60°C water bath; bench centrifuge for Eppendorf tubes; liquid scintillation counter; equipment and chemicals for gel electrophoresis (page); lyophyliser.

2.2 Pilot experiment to determine the optimal reaction time to add 15-20 cytidylates to a single stranded, chemically prepared oligodeoxynucleotide (n = 12-20)

Reaction mixture

10 µL	1	M	cacodylate-K, pH 7.2
10 µL	20	mM	$MgCl_2$
5 µL	1	mM	dCTP
5 µL			$[\alpha\text{-}^{32}P]$dCTP 10^7cpm
5 µL	1	mM	DTT
10 µL	0.1	mM	$(pN)_{12-20}$
4 µL			H_2O
1 µL	14 units		TdT

Incubate the reaction mixture at 37°C for 5 min. Take out two 2 µL aliquots, apply to GF/A filter. Wash one filter with 10% TCA and count both filters by Cerenkov counting (gives total cpm and minus enzyme blanc). Put sample on ice for 10 min, add 1 µL TdT and continue incubation at 15°C. Take out 2 µL aliquots at 0, 2, 5, 7, and 10 min. Prior to counting wash with 10% TCA.

From the pmol dCTP in the reaction mixture (5000) and the total radioactivity calculate the cpm per pmol dCTP. Plot time versus cpm (substract minus enzyme blanc) and calculate the number of dC molecules attached to the oligodeoxynucleotide. From the graph determine the incubation time required for 20 nucleotide addition.

2.3 Preparative scale experiment

For an experiment on a preparative scale the reaction mixture is increased 5fold. Incubate at 15°C for the time determined in the previous experiment. Take two 2 µL aliquots and check for incorporation. Calculate the average number of nucleotide residues added to the oligodeoxynucleotide as outlined above. Stop the reaction by adding 5 µL 4 M NaCl and 3 µL 0.2 M EDTA Na_2 and heat to 60-70°C for 5 min. Next the solution is extracted with chloroform:isoamyl acetate (100:1, v/v) and the organic phase is reextracted with water. The aqueous phases are combined and the oligonucleotide is precipitated by the addition of 2 volumes ethanol. After storage for 16 h at -20°C the tailed oligonucleotide is isolated by centrifugation, the pellet is dissolved in a small volume of water and lyophylized.

2.4 Tailing of linearised plasmid pBR322 with guanylates

2.4.1 Linearisation of the plasmid with Bam H1 (see Gatz and Hillen)

10 µg pBR322 DNA is digested with 50 units Bam H1 in 25 µL restriction buffer for 1 h at 37°C using a total volume of 250 µL.
Restriction buffer

1 M Tris-HCl, pH 7.2
50 mM $MgCl_2$
20 mM ß-mercaptoethanol

The solution is put on ice and the efficiency of hydrolysis is checked by page (see Klock). The linearised plasmid, more than 80%, migrates shorter (about 1 cm on a 20 cm gel) than the supercoiled circular plasmid.

The reaction is stopped by the addition of 10 µL 0.5 M EDTA, 5 µL 20% SDS and heating at 55°C for 5 min. The mixture is extracted twice with an equal volume of phenol:chloroform (1:1, v/v). The aqueous phase is saved, adjusted to 0.2 M NaOAc, and mixed with 2.5 volumes of ethanol. After 2 h at -20°C the precipitated DNA is sedimented by centrifugation at 10 000 rpm for 15 min. The supernatant is discarded and the DNA is dissolved in 10 mM Tris-HCl, pH 8.3 (100 µL).

2.4.2 Tailing of the linearised plasmid with oligoguanylate

Procedure

The following solutions are added in order

20 µL	0.7 M	cacodylate-K, pH 7.6
41 µL		H_2O
10 µL	10 mM	$CoCl_2$
5 µL	2 mM	DTT
1 µL	10 mM	dGTP
2 µL		[α-^{32}P]dGTP 10^7 cpm
20 µL	~0.4 pmol	lin. pBR322 DNA
1 µL	14 units	TdT

The mixture is incubated at 37°C for 5 min, then chilled in ice for 10 min. 14 units of TdT are added and the solution is incubated for 30 min at 15°C. The addition of [^{32}P]dGMP residues to the DNA is monitored by sampling 2 µL aliquots at 0, 2, 5, 10, and 30 min. The aliquots are spotted on GF/A filters, the filters are washed with 10% TCA and counted. When the incorporation reaches a plateau, usually 15-20 dGMP residues incorporated, the enzyme is inactivated by adding 12 μL 4 M NaCl and 2 μL 0.2 M EDTA-Na_2 and heating to 60-70°C for 5 min.

Calculate the number of dGMP residues added to the plasmid, and use the incubation time, which results in 15-20 nucleotides added, for the experiment on a preparative scale. The tailed plasmid is extracted with chloroform:isoamylalcohol and ethanol precipitated as described before.

REFERENCES

[1] Bollum, F.J. (1974) The Enzymes 10, 145

[2] Krakow, J.S., Coutsogeorgopoulos, C., and Cannellakes, E.S. (1961) Biophys. Biochem. Res. Comm. 5, 477

[3] Maxam, A.M. and Gilbert, W. (1980) Methods in Enzymology 65, 499

[4] Roychoudhury, R. and Kössel, H. (1971) Eur. J. Biochem. 22, 310

[5] Roychoudhury, R. (1972) J. Biol. Chem. 247, 3910

[6] Padmanabhan, R., Wu, R., and Calendar, R. (1974) J. Biol. Chem. 249, 6197

[7] Sehiya, T., van Ormondt, H., and Khorana, H.G. (1974) J. Biol. Chem. 250, 1087

[8] Kössel, H. and Roychoudhury, R. (1971) Eur. J. Biochem. 22, 271

[9] Bertazzoni, U., Ehrlich, S.D., and Bernardi, G. (1974) Methods in Enzymology 29E, 355

[10] Roychoudhury, R., Fischer, D., and Kössel, H. (1971) Biochem. Biophys. Res. Commun. 45, 430

[11] Kössel, H., Roychoudhury, R., Fischer, D., and Otto, A. (1974) Methods in Enzymology 29E, 322

[12] Wu, R., Tu, C.D., and Padmanabhan, R. (1973) Biochem. Biophys. Res. Commun. 55, 1092

[13] Jay, E., Bambara, R., Padmanabhan, R., and Wu, R. (1974) Nucl. Acids Res. 1, 331

[14] Morrison, M.R. and Griffin, W.S.T. (1981) Ann. Biochem. 113, 318

[15] Roychoudhury, R., Jay, E., and Wu, R. (1976) Nucl. Acids Res. 3, 863

LIGATION OF DNA FRAGMENTS

Gerd Klock

Institut für Organische Chemie und Biochemie
Technische Hochschule Darmstadt

SUMMARY

The joining of blunt-ended or sticky-ended DNA fragments with T4 DNA ligase is described.

1 INTRODUCTION

T4 DNA ligase (EC 6.5.1.3) is used to join DNA fragments which are either blunt-ended or contain complementary sticky ends. The action of T4 DNA ligase requires a 5'-terminal phosphate on the donor end and a free 3'-terminal hydroxyl group on the acceptor DNA sequence [1,2]. The ligase catalysed joining of DNA ends is the crucial step in recombinant DNA technology. It has also been used to study the flexibility of DNA by determining the degree of circularisation which occurs for a DNA fragment of given length compared to the bimolecular dimerisation of two DNA sequences [3]. This protocol describes the determination of optimal conditions for sticky- and blunt-ended ligation. These conditions then may be used to join chemically synthesised DNAs with various sequences and chain lengths.

2 EXPERIMENTAL SECTION

Materials

T4 DNA ligase; purified DNA fragments suitable for blunt-end and sticky-end ligation; acrylamide gels

Buffers:

2 L staining solution

5 µg ethidium bromide per liter distilled water

1 mL T4 DNA ligase buffer (10X)

4	mM ATP
660	mM Tris-HCl, pH 7.6
66	mM $MgCl_2$
100	mM dithiothreitol

1 mL "blue juice"

1	%	Na-dodecylsulfate
25	%	glycerol
0.08	%	bromophenol blue
40	mM	Tris-base
10	mM	Na-acetate
2	mM	EDTA

adjusted to pH 8.3 with glacial acetic acid

1 L electrophoresis buffer (5X)

- 540 g/L Tris-base
- 36.5 g/L EDTA
- 10 g/L NaOH
- 275 g/L boric acid

adjusted to approximately pH 8.5 with NaOH

Equipment:

Gel electrophoresis apparatus for 20 cm x 20 cm slab gels, 2 mm thick. Photography equipment for ethidium bromide stained gels.

Procedure:

Time course of the ligation reaction

1.5 μg of purified DNA fragments, 245-bp in length, containing one Eco RI end and one Hae III end is incubated in a total volume of 60 μL

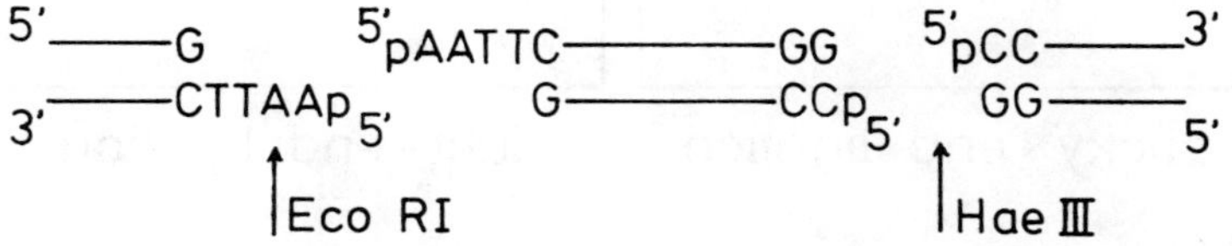

```
5'——G         5'pAATTC————GG    5'pCC————3'
3'——CTTAAp5'    G————CCp5'    GG————5'
    ↑Eco RI                  ↑Hae III
```

ligase buffer, containing 0.12 units T4 DNA ligase for sticky end ligation and 6 units T4 DNA ligase for blunt end ligation at 20°C. The course of the reaction is followed by withdrawing 8 μL (0.2 μg DNA) samples after 5, 10, 15, 25, 40, and 60 min. The ligation reaction is terminated by the addition of 10 μL "blue juice" to each sample with incubation for 2 min at 70°C. All the samples together with the starting material and DNA fragments generated by a Hae III digest of pBR322 as size markers are then subjected to electrophoresis on a 5% polyacrylamide gel containing 10% glycerol. The gel is run at 150 V for 5 h, immersed in 0.05% ethidium bromide for 10 min. The gel is photographed under uv-irradiation and the intensity of the bands is estimated from this picture (Figure 1). The minimal time needed for completion of the reaction is determined and may be used for a preparative ligation.

For preparative purposes less enzyme and a prolonged incubation time (16 h) may be used, to reduce the amount of the expensive enzyme. For transformation experiments the ligated DNA fragments may be used without further purification. However, if subsequent phosphorylation with polynucleotide kinase is required, the oligonucleotide should be purified by page, hplc or by chromatography on RPC-5 [4].

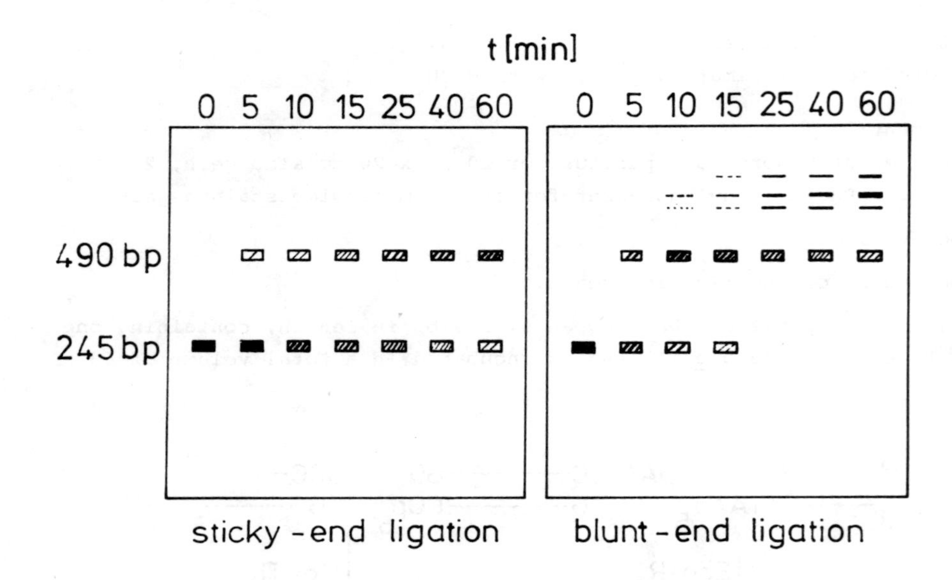

Figure 1 Schematic drawing of the page separation of the ligated oligonucleotides. The intensity of bands is indicated by the hatching in the figure.

REFERENCES

[1] Weiss, B. et al. (1968) J. Biol. Chem. 243, 4543

[2] Heinecker, H.C. et al. (1976) Nature 263, 748

[3] Shore, D., Langowski, J., and Baldwin, R.L. (1981) Proc. Natl. Acad. Sci. 78, 4833

[4] Hillen, W., Klein, R.D., and Wells, R.D. (1981) Biochemistry 20, 3748

CONSTRUCTION OF RECOMBINANT PLASMIDS USING PREPURIFIED DNA FRAGMENTS

Christiane Gatz and Wolfgang Hillen

Institut für Organische Chemie und Biochemie
Technische Hochschule Darmstadt

SUMMARY

This experiment introduces the basic techniques for the insertion of DNA into plasmids, the transformation of recombinant plasmids to E. coli and the selection and screening of the resulting clones.

1 INTRODUCTION

The combination of chemical synthesis of DNA sequences with cloning procedures may serve several purposes. Three important aspects will be briefly considered here: a) functional characterization of DNA *in vivo*, b) purification of specific sequences from mixtures of fragments, and c) amplification of large DNA fragments for preparative purposes.

The chemical synthesis of DNA focuses on regulatory sequences as well as structural information for proteins. In order to test the *in vivo* functions or to express structural information *in vivo*, these sequences have to be introduced into a host cell. Plasmids provide convenient vehicles for this purpose, because they generally allow a drug resistance selection for their presence and, in some cases, double selections for the screening of inserted DNA [1]. E. coli has been mainly used so far as the host to express chemically synthesized genetic information [2,3]. The insertion of promotor sequences upstream of the β-galactosidase gene can also be used to estimate the efficiency of this regulatory sequence in initiating transcription. Similar approaches may also be feasible for termination signals.

The second aspect to be considered makes use of the fact that a clone of a recombinant strain will contain a single plasmid. Thus, if the chemical synthesis of especially large DNA fragments yields a mixture of sequence isomers or other impurities which are difficult to remove, then a cloning experiment may be a convenient strategy to select for particular sequences. These sequences will then be contained in a recombinant plasmid and may be subjected to further analysis in this state, or the construction of recombinant plasmids can be designed such that it permits purification of the inserted DNA from the plasmid (compare figure below).

The last aspect to be considered is the procedure to construct and use plasmids for the purpose of large scale preparation of DNA fragments. The combination of chemical synthesis and subsequent amplification of the sequences *in vivo* could be the strategy of choice for rather large DNAs. In that case chemical synthesis on a large scale may still be laborious, but the yield of large fragments from amplifiable plasmids is excellent [4,5]. For preparative purposes the inserted DNA has to be located between single restriction sites. This goal can be achieved either by the use of linker fragments or by the reconstruction of restriction sites [6]. If desired, oligo-insertion plasmids can then be constructed to increase the ratio of insert to vector DNA [6,7].

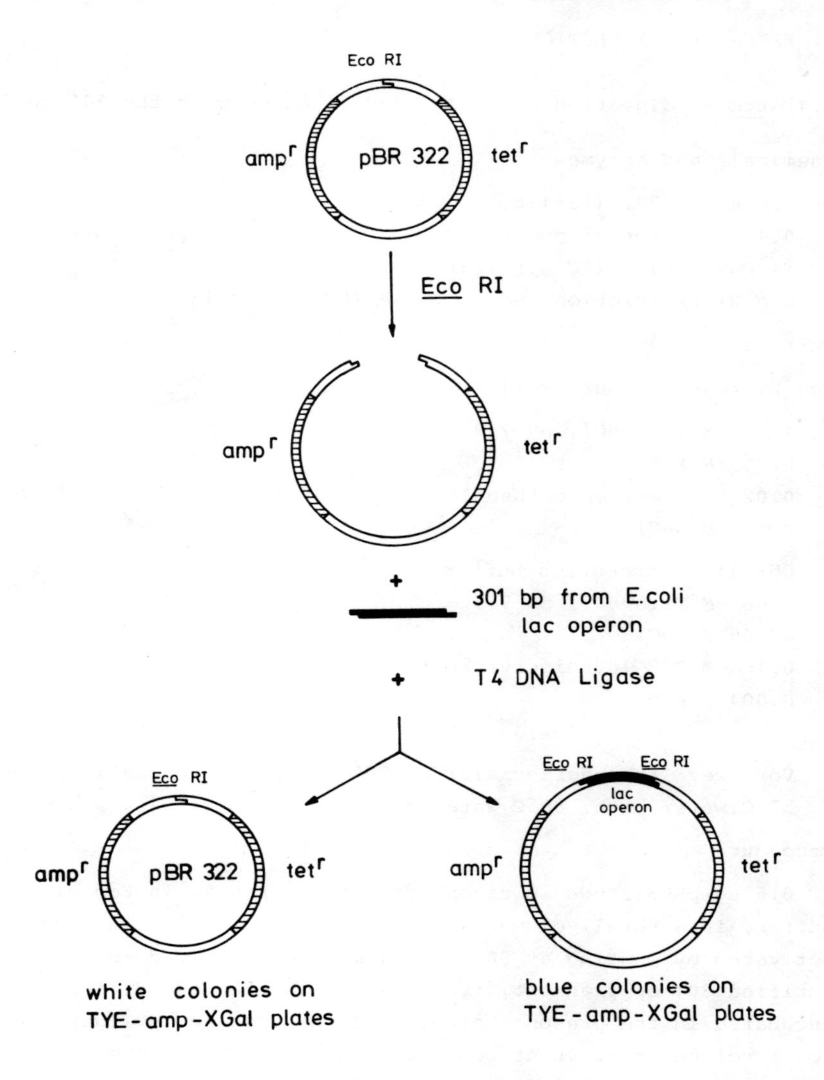

The following experiment is designed to demonstrate the procedures involved in the insertion of a DNA fragment into a plasmid vehicle, and the selection and screening of recombinants. These methods are demonstrated by cloning a 301-bp DNA fragment originating from the E. coli lactose control region into the Eco RI site of pBR322 [1]. Prior to this experiment the original Hae III ends of the 301-bp DNA were converted to Eco RI ends [7]. This DNA was particularly chosen because it allows rapid selection and screening [6].

2 EXPERIMENTAL SECTION

2.1 *Eco* RI digestion of pBR322 and ligation with the 301-bp fragment

Chemicals and enzymes

0.5 µg pBR322 (4361-bp)
0.1 µg 301-bp fragment
T4 DNA ligase (EC 6.5.1.3)
Eco RI restriction endonuclease (EC 3.1.23.13)

Buffers

Eco RI reaction buffer (10X)

1 M Tris-HCl, pH 7.2
0.05 M $MgCl_2$
0.02 M ß-mercaptoethanol
0.5 M NaCl

T4 DNA ligase reaction buffer (10X)

0.66 M Tris-HCl, pH 7.5
0.066 M $MgCl_2$
0.1 M DTE (dithioerythrol)
0.004 M ATP

Materials

Vortexer; Eppendorf-vials; 20/200 µL Gilson automatic pipettes; 37°C water bath; 80°C water bath

Procedure

0.5 µg pBR322 DNA is digested with 8 U *Eco* RI in *Eco* RI reaction buffer, in a total volume of 30 µL for 1 h at 37°C. The enzyme is inactivated by heating at 80°C for 1 min. A threefold molar excess of purified 301-bp fragment, i.e. 0.1 µg, is then added. This mixture is incubated in the presence of 0.1 U T4 DNA ligase in ligase buffer in a total volume of 50 µL at room temperature for 1 h. The ligation solution is directly used for the transformation experiment.

2.2 Transformation and selection of the recombinant plasmids

Media and buffers

L-Broth:	10	g/L Bacto Tryptone
	5	g/L Yeast Extract
	10	g/L NaCl
TYE-plates:	15	g/L Agar
	8	g/L NaCl

10 g/L Bacto Tryptone
5 g/L Yeast Extract
40 mg/L ampicillin
40 mg/L 5-bromo-4-chloro-3-indolyl-ß-galactoside (XGal) dissolved in 1 mL dimethylformamide (DMF)

0.1 M $CaCl_2$ (sterile)

SSC-buffer: 8.76 g/L NaCl
4.41 g/L Na_3citrate
adjusted to pH 7.0 with HCl

SSC:0.1 M $CaCl_2$ (1:2, v/v)

Materials

sterile 30 mL tubes; sterile pipettes (10, 5, 2, 1 mL); Sorvall centrifuge vials; Rotor SS 34; shaking equipment; 37°C water bath; plating equipment

Procedure

1 mL overnight culture of the recipient cells (E. coli C600) grown in L-Broth is transferred to 20 mL L-Broth and shaken at 37°C until the optical density is 0.6 A_{650} (early log phase). The culture is chilled for 10 min, centrifuged at 8000 rpm in a Sorvall, Rotor SS34 at 0°C for 10 min and resuspended in 10 mL sterile 0.1 M $CaCl_2$. The cells are held at 0°C for 20 min and then centrifugation is repeated as above. The cells are then resuspended in 1/10 the original volume 0.1 M $CaCl_2$ and reincubated at 0°C for 15 min. 200 µL of the now competent cell suspension is transferred to 30 mL tubes containing 5 and 25 µL of the ligation mixture from experiment 2.1 and 100 µL SSC:0.1 M $CaCl_2$ (1:2, v/v) under sterile conditions. This mixture is incubated for 3o sec at 37°C followed by 90 min at 0°C. The tubes should be shaken every 15 min. 4 mL L-Broth are added under sterile conditions and the cells are allowed to grow out while shaking at 37°C for 90 min. 0.2 mL of the out growth is plated on TYE plates containing 40 mg/L ampicillin and 40 mg/L XGal. After 20 h of incubation at 37°C, colonies containing recombinant plasmids with the 301-bp sequence inserted in pBR322, can be identified by their blue colour. Between 30 to 50% of the colonies should contain recombinant plasmids in this experiment.

REFERENCES

[1] Bolivar, F., Rodriguez, R.L., Greene, P.J., Betlach, M.C., Heynecker, H.C., Boyer, H.W., Crosa, J.H., and Falkow, S. (1977) Gene 2, 95

[2] Goeddel, D.V., Kleid, D.G., Bolivar, F., Heynecker, H.C., Yansary, D.G., Crea, R., Hirose, T., Kraszewski, A., Ithakura, K., and Riggs, A.D. (1979) Proc. Natl. Acad. Sci. USA 75, 3727

[3] Edge, M.D., Green, A.R., Heathcliffe, G.R., Meacoch, P.A., Schuch, W., Scanlon, D.B., Atkinson, T.C., Newton, C.R., and Markham, A.F. (1981) Nature 292, 756

[4] Hardies, S.C. and Wells, R.D. (1979) Gene 7, 1

[5] Hillen, W., Klein, R.D., and Wells, R.D. (1981) Biochemistry 20, 3748

[6] Hardies, S.C., Patient, R.K., Klein, R.D., Ho, F., Reznikoff, W.S., and Wells, R.D. (1979) J. Biol. Chem. 254, 5527

[7] Hardies, S.C., Hillen, W., Goodman, T.C., and Wells, R.D. (1979) J. Biol. Chem. 254, 10128

PROPOSED STRATEGY FOR LARGE SCALE PRODUCTION OF DNA SEQUENCES USING CHEMICAL AND BIOLOGICAL METHODS

Wolfgang Hillen

Institut für Organische Chemie und Biochemie
Technische Hochschule Darmstadt

SUMMARY

A strategy for large scale production of a short DNA fragment is proposed. The construction scheme of the DNA involves chemical synthesis of three dodecanucleotides, their enzymatic phosphorylation and ligation via complementary protruding ends, the filling in of remaining single stranded ends and finally the construction of recombinant plasmids containing several copies of the desired sequence between Ava I and Hpa I sites. This plasmid represents a storage form of the target DNA which may be prepared in large amounts from a transformed E. coli strain. The principle of this scheme is to synthesize a master copy of the desired sequence which is then reproduced by bacteria.

1 INTRODUCTION

Large amounts of a biologically active DNA fragments are desirable for physico-chemical investigations as nmr-, laser raman-, and X-ray diffraction spectroscopy [1-3]. This task is of general interest because these studies can yield insights for example in the specificity of replication, regulation of gene expression, or other fundamental biological reactions on a molecular basis [4,5]. Therefore, protocols have been described to prepare biologically important DNA sites on small fragments [6-8]. However, target sites for proteins which recognize specific DNA sequences are usually short and the observation of specific effects is often hampered by the presence of excess flanking non-specific DNA.

As outlined in the previous chapters chemical synthesis of DNA is a powerful tool to obtain specific sequences. Especially the synthesis of short DNAs is straightforward and often advantageous over the isolation from natural DNA sources. On the other hand, the reproduction of a given DNA sequence in large amounts is very efficiently achieved by E. coli leaving no problem arising from incomplete deprotection of synthetic DNA or the presence of sequence isomers in the sample etc. Also, the construction of a suitable plasmid aids to facilitate the final purification of the desired DNA [6-8]. On this basis it seems to be a reasonable approach to large scale production of a short, specific DNA sequence to design and chemically synthesize the master copy and then use recombinant methods to program an E. coli strain for the efficient copying of this DNA.

The following protocol is designed as a theoretical concept to approach the large scale preparation of the tet operator sequence from the transposon Tn10 [9,10]. It involves the chemical synthesis of three DNA dodecanucleotides, their phosphorylation, ligation, the "fill in" of protruding single stranded ends, and finally the construction of a multi-insertion plasmid to be transformed into the E. coli production strain.

2 OVERVIEW OF THE CONSTRUCTION SCHEME

This scheme is designed to serve the following purposes: i) The tet operator palindrom shall be accessible to preparation with as little as possible flanking DNA; ii) The resulting sequence shall be blunt ended; iii) The chemically synthesized DNA shall be suitable for the construction of multi-insertion plasmids. The sequence of the tet operator palindrom communicated to us by K. Bertrand and W.S. Reznikoff is:

5'ACTCTATCATTGATAGAGT3'
3'TGAGATAGTAACTATCTCA5'.

It is obvious that the end nucleotides fit into recognition sites for the restriction endonuclease Hpa I (Hin cII) which is 5'GTT|AAC3'
3'CAA|TTG5'.

This enzyme produces a blunt ended cut as indicated and the last two nucleotides overlap with the tet operator palindrom. Thus, the attachment of a complete Hpa I site on both ends of the operator will result in a blunt ended DNA fragment 21 bp long with just one additional bp on each end.

The construction of multi-insertion plasmids is limited by the observation that tandem copies of the same sequence in the head to head or tail to tail orientation tend to be unstable in E. coli [8]. Because DNA ligase does not select for orientation the probability of parallel orientation drops sharply with the number of ligation events. In order to introduce a selection for head to tail joining in the ligation reaction one can make use of the non-equivalent protruding ends produced by the restriction endonuclease Ava I [11]. The recognition site for Ava I is 5'C↓CCGAG3' in pBR322 with the indicated cleaving sites [12].
3'GGGCT↑C5'

A DNA restriction fragment with these Ava I protruding ends gives rise to preferred head to tail ligation because head to head or tail to tail joining creates AC and GT mismatches, respectively [11]. In approximating the selective advantage from the mismatch enthalpy one obtains for the probability of an all head to tail ligated DNA from n monomers without selection $(1/2)^{n-1}$ whereas the Ava I sites would give roughly $(3/4)^{n-1}$. For a 10 mer the numbers are one all parallel orientation in 512 oligomers without and one in 13 with selection. For a 20 mer the probabilities would be one in 5.2×10^5 and one in 237. This rough approximation demonstrates very clearly the advantage of reconstructing non-equivalent Ava I sites over any site with equivalent protruding ends [11]. The complete construction scheme is summarized in figure 1.

3 CONSTRUCTION OF THE DESIRED SEQUENCE

3.1 Chemical synthesis

Figure 2 shows the three dodecanucleotides to be synthesized chemically. The 12 mer A can be used twice because the central AT base pair of the final palindrom is not included in this sequence. Owing to the fact that the symmetry plane of this DNA is centered in the AT base pair rather than between two base pairs the dodecanucleotides B and C

Ⓐ

5' ACTCTATCATTGATAGAGT 3'
3' TGAGATAGTAACTATCTCA 5'

desired sequence

Ⓑ

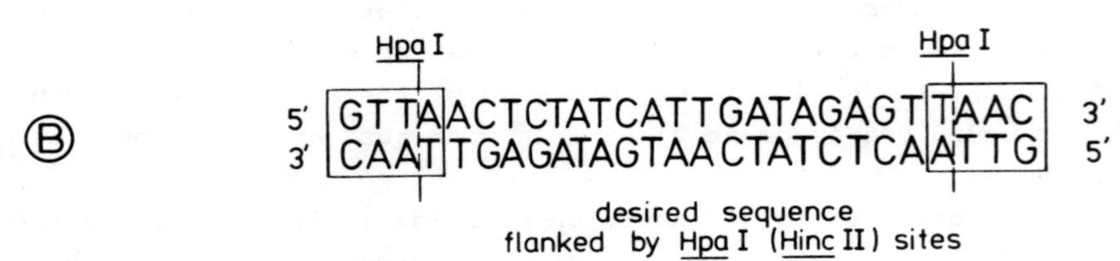

desired sequence
flanked by Hpa I (Hinc II) sites

Ⓒ

5' TCGGGTTAACTCTATCATTGATAGAGTTAAC 3'
3' CAATTGAGATAGTAACTATCTCAATTGAGCC

Ava I protruding ends added
for oligomerisation

Figure 1 Construction scheme of the desired sequence. Panel A shows the sequence of the tet operator. In Panel B the two Hpa I restriction sites, which allow the blunt ended excision of this tet operator are added. The cutting specificity of Hpa I is indicated by dashed vertical lines. Panel C displays the final sequence containing the Ava I protruding ends

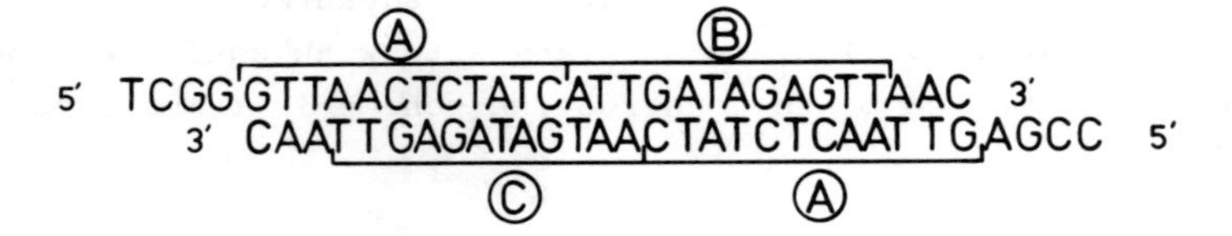

Figure 2 Oligonucleotides to be chemically synthesized. The dodecamers A, B, and C are synthesized using chemical methods. The remaining nucleotides are added afterwards by enzymatic procedures

differ in only one nucleotide at the position 11 (counted from the 3'-end as usual in chemical DNA synthesis). The latter two dodecamers may be cosynthesized either by splitting the sample for step eleven or by adding T plus A in step eleven to yield the mixed sample.

3.2 Phosphorylation of the dodecamers

Prior to the phosphorylation the dodecamers have to be deprotected and partially purified. The latter is usually done by hplc, however, we have run the oligonucleotides on 20% polyacrylamide gels and eluted the respective bands. Ethanol precipitation can then be performed to remove contaminations from the gel. It should be noted though, that the precipitation does not succeed with all oligonucleotides. We have used the gel eluted samples successfully without further purification when the purest acrylamide available was used for the gels. The phosphorylation succeeds with the aid of T4 polynucleotide kinase and [γ-^{32}P]ATP. Separation of the phosphorylated dodecamer from the remaining ATP is done on 20% polyacrylamide gels and the reaction product is eluted from the gel.

3.3 Ligation of the double strands

Figure 3 shows the principle of the ligation reaction which is done with T4 DNA ligase (described by Klock, this volume). The middle protruding ends in figure 3 are base complementary whereas the outside ends are not. This should direct the ligation reaction to yield the desired products.

3.4 "Filling in" of the protruding ends

The reaction to fill in the protruding single stranded ends to yield a blunt ended double strand is also shown in figure 3. Aside from DNA polymerase I the Klenow fragment or T4 DNA polymerase [13] may be used to add the complementary nucleotides.

3.5 Construction of the *Ava* I protruding ends

Figure 4 explains the experimental approach to add the four single stranded nucleotides otherwise produced by cleavage with the restriction endonuclease *Ava* I on each 5'-end of the DNA. This approach involves the cleavage of pBR322 at its single *Ava* I site and the filling in of the resulting single stranded ends. Figure 4 demonstrates clear-

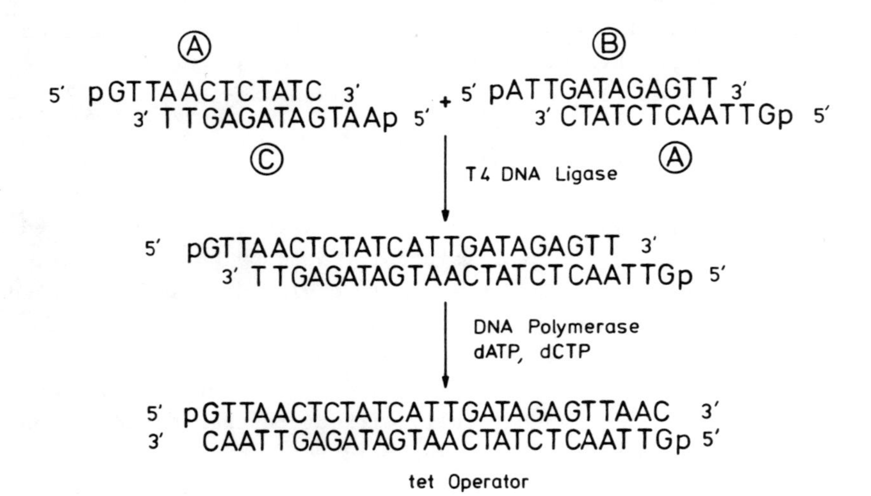

Figure 3 Joining of the dodecamers. After phosphorylation the dodecamers are ligated using their complementary protruding ends and the remaining protruding ends are filled in using DNA polymerase. The result is the fragment B indicated in figure 1

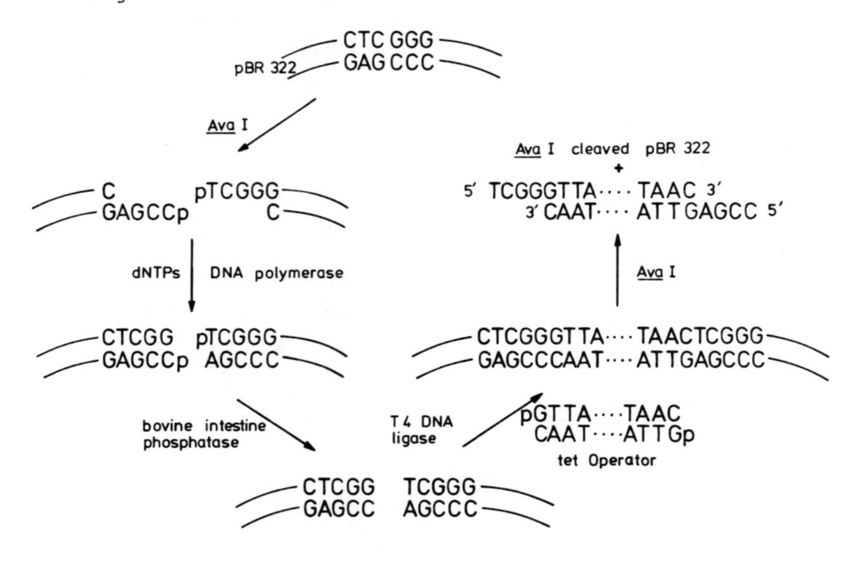

Figure 4 Strategy to add the <u>Ava</u> I protruding ends. Details of this figure are described in the text

ly that the ligation of the synthetic DNA into the filled in Ava I site of pBR322 regenerates Ava I sites on both ends of the fragment. Because there is no screen for the presence of the inserted DNA on the plasmid the filled in pBR322 is dephosphorylated using calf intestine phosphatase prior to the ligation. This prevents recircularisation of the vector and should make sure, that about 95% of the selected colonies are recombinants. The efficiency of the procedure may be checked by inserting DNA with an easy screen as the lac operator DNA. After the ligation the reaction mixture is used to transform E. coli C600 (see Gatz and Hillen, this volume). Preparation and Ava I digestion of the recombinant plasmid will yield the Ava I "sticky" ends on the synthetic DNA.

3.6 Construction of multi insertion plasmids

Two different approaches are possible to arrive at multiple insertion plasmids. The first is to oligomerize the synthetic DNA via their Ava I sticky ends, separate the oligomers on polyacrylamide gels, elute the desired bands and ligate them into Ava I cleaved pBR322. The second involves just the ligation of Ava I cleaved pBR322 with an excess of synthetic DNA. Due to the lack of a screen and the small size of the DNA, the first design seems to be more efficient, especially when the terminal phosphates are removed from pBR322 as described above.

4 CONCLUSION

This synthetic scheme is as yet a proposed construction which is included in this volume to demonstrate the possible efficiency of combining chemical, enzymatic, and biological methods. The DNAs A, B, and C have been synthesized during the course and were subsequently phosphorylated and ligated. The remaining steps are currently under evaluation.

REFERENCES

[1] Early, T.A., Kearns, D.R., Hillen, W., and Wells, R.D. (1981) Biochemistry 20, 3764

[2] Early, T.A., Kearns, D.R., Hillen, W., and Wells, R.D. (1981) Biochemistry 20, 3756

[3] Wang, A.H.-J., Quigley, G.J., Crawford, J.L., van Boom, J.H., von der Marel, G., and Rich, A. (1979) Nature 282, 680

[4] Wells, R.D., Goodman, T.C., Hillen, W., Horn, G.T., Klein, R.D., Larson, J.E., Müller, U.R., Neuendorf, S.K., Panayotatos, N., and Stirdivant, S.M. (1980) Progr. Nucl. Acids Res. Mol. Biol. 24, 167

[5] von Hippel, P.H. in: Biological Regulation and Development (R.F. Goldberger ed.) Vol. 1, p 279, Plenum Press, New York 1979

[6] Hillen, W., Klein, R.D., and Wells, R.D. (1981) Biochemistry 20, 3748

[7] Sadler, J.R., Tecklenberg, M., and Betz, J.L. (1979) Gene 8, 279

[8] Kallai, O.B., Rosenberg, J.M., Kopka, M.L., Takano, T., Dickerson, R.E., Kan, J., and Riggs, A.D. (1980) Biochim. Biophys. Acta 606, 113

[9] Wray, L.V., Jorgenson, R.A., and Reznikoff, W.S. (1981) Bacteriol. 147, 297

[10] Hillen, W., Klock, G., Kaffenberger, I., Wray, L.V., and Reznikoff, W.S. (1982) J. Biol. Chem., in press

[11] Hartley, J.L. and Gregori, T.J. (1981) Gene 13, 347

[12] Sutcliffe, J.G. (1978) Nucl. Acids Res. 5, 2721

[13] Wartell, R.M. and Reznikoff, W.S. (1980) Gene 9, 307

POLYNUCLEOTIDE PHOSPHORYLASE CATALYSED SYNTHESIS OF OLIGONUCLEOTIDES OF DEFINED SEQUENCE

Anne Lang and Hans Günter Gassen

Institut für Organische Chemie und Biochemie
Technische Hochschule Darmstadt

SUMMARY

Polynucleotide phosphorylase from either M. luteus or E. coli is used for the synthesis of oligoribonucleotides and polyribonucleotides. The enzyme is made primer dependent by limited hydrolysis with trypsin. The method for the synthesis of trinucleotides starting from the dinucleotide is outlined. Hydrolytic enzymes such as ribonuclease T1 are used to obtain the trinucleotides in high yields. Furthermore, the procedures for the synthesis of block oligonucleotides such as $AUG(U)_n$ and polynucleotides are given.

1 INTRODUCTION

Oligo- and polynucleotides of defined sequence are required as model compounds for studies on the mechanism of protein synthesis and for the examination of interactions between nucleic acids and nucleic acids and proteins. Although a number of chemical and enzymatic methods for the synthesis of oligonucleotides are known (see previous sections of this manual), polynucleotide phosphorylase (PNPase) has widely been used for the preparation of polynucleotides and oligonucleotides of defined sequence, since it does not require any DNA or RNA template. The preparation of highly purified and of primer dependent polynucleotide phosphorylase has been described [1,2].

Several other methods for its preparation have been reported [3,4], requiring less time to prepare an enzyme well suited for preparative purposes [3]. The source for the enzyme is mainly M. luteus or E. coli, although PNPase from a great variety of organisms has been isolated.

Polynucleotide phosphorylase can be converted to the primer dependent form ($PNPase^{p}$) by limited digestion with trypsin [2,3]. An alternative method for the preparation of $PNPase^{p}$ has been described by Letendre and Singer [5], which involves reduction of the enzyme with dithioerythritol in 1 M guanidinium chloride and subsequent treatment with N-ethylmaleimide. Primer dependent polynucleotide phosphorylase is unable to catalyse the de novo synthesis of polynucleotides from nucleoside 5'-diphosphate but requires an oligonucleotide primer. The primer will function as the starting molecule for the growing polynucleotide chain.

The chain length of the oligonucleotides to be synthesised by primer dependent polynucleotide phosphorylase can be controlled in two different ways [6].

a) The "kinetic method"

Increasing NaCl concentrations in the incubation mixture result in a decrease of the reaction rate by which nucleotidyl residues are added to the growing oligonucleotide chain. At a given NaCl concentration, chain elongation will practically stop at a certain chain length of the polymer. Thus it is possible to control chain length by NaCl concentration [7]. Incubation times range from 1 to 6 h.

b) The "equilibrium method" [8]

If the polymerisation reaction catalysed by $PNPase^{p}$ is allowed to reach thermodynamic equilibrium, chain length distribution of the products is determined by the ratio of primer to nucleoside-5'-diphosphate initially present in the reaction mixture. Incubation times range from 24 to 72 h.

Under the reaction conditions used in the equilibrium method PNPase catalyses the reverse reaction as well, i.e. phosphorolysis of oligonucleotides and polynucleotides except of the 5'-terminal dinucleotide. Thus only dinucleoside phosphates can be used as primers in this reaction. Oligonucleotides longer than two residues will be partially phosphorolysed. The yields in the synthesis of trinucleotide diphosphates depend on the nature of the base moiety, usually they are in the range of 20-30%, poor yields are obtained with ppG [6].

Addition of RNase T_1 or RNase A to the reaction mixture yields oligonucleotides with 3'-terminal guanosine-3'-phosphate in the case of RNase T_1 and oligonucleotides with 3'-terminal pyrimidine nucleoside--3'-phosphate with RNase A (the primers of course should not contain these nucleosides). For removal of the 3'-terminal phosphate group the reaction mixture has to be treated with phosphatase. With this modification trinucleoside diphosphates can be obtained in high yields [9,10]. Low yields are obtained if pyrimidine oligonucleotide primers and purine nucleoside diphosphates are used [11]. Increased primer concentration and high enzyme concentrations have been applied to overcome this disadvantage [12].

A different way to obtain oligonucleotides of defined sequence is the use of primer independent PNPase and 2'- or 3'-substituted nucleoside-5'-diphosphates as substrates. However, careful analysis of this reaction has shown, that phosphorolysis of primers containing more than 3 nucleotides is not neglegible [13,14]. Sninsky et al. [15] have reported the use of an auxiliary enzymatic system, that removes inorganic phosphate from the reaction mixture. This phosphate trapping system should prevent inorganic phosphate from being used in the phosphorolytic cleavage of oligonucleotides by PNPase. However, both methods did not gain wide acceptance.

Oligonucleotides prepared by polynucleotide phosphorylase, especially trinucleotides and tetranucleotides are used as donors and acceptors in the RNA ligase catalysed synthesis of RNA fragments (see Uhlenbeck, Ohtsuka and Eckert).

2 EXPERIMENTAL SECTION

2.1 Materials and equipment

Polynucleotide phosphorylase (EC 2.7.7.8) prepared as to [3] from M. luteus or from E. coli or obtained from commercial sources; ribonuclease T1 (EC 3.1.27.3); alkaline phosphatase (EC 3.1.3.1) from calf intestine, free of 6-aminodeaminase; trypsin (EC 3.4.21.4); soybean trypsin inhibitor; ADP, CDP, GTP, UDP; [^{14}C]ADP; ApA; ApU; Tris base;

$MgCl_2$; EDTA-Na_2; NaCl; NH_4OAc; HCl; 1 M $(C_2H_5)_3NH \cdot HCO_3$; methanol; ethanol; phenol saturated with H_2O; toluene containing 0.4% diphenyloxazol; DEAE-cellulose; QAE Sephadex; Dowex 50[NH_4^+] 100 mesh; Sephadex G-25 fine; Biogel P2; polyethylene tubes, 400 µL and 2 µL; microliter test plates with 15 µL wells; small tank for paper chromatography; paper strips (1 cm x 11 cm) cut from chromatography paper Schleicher and Schüll 2316; equipment for liquid column chromatography; liquid scintillation counter.

2.2 Preparation of primer dependent polynucleotide phosphorylase [3]

Reagents

1 mL	5 mg/mL	polynucleotide phosphorylase 500 U/mL
100 µL	1 mg/mL	trypsin
100 µL	1 mg/mL	soybean trypsin inhibitor
100 µL	5 M	NaCl
1 mL		mix for primer independent polymerisation

Polymerisation mix (-primer) (1 mL)

140 µL	1 M	Tris-HCl, pH 9.5
10 µL	1 M	$MgCl_2$
10 µL	40 mM	EDTA-K
300 µL	160 mM	[^{14}C]ADP 67 $mCi \cdot mol^{-1}$
540 µL		H_2O

Polymerisation mix (+primer) (1 mL)

140 µL	1 M	Tris-HCl, pH 9.5
10 µL	1 M	$MgCl_2$
10 µL	40 mM	EDTA-K
300 µL	160 mM	[^{14}C]ADP 67 $mCi \cdot mol^{-1}$
40 µL	35 mM	ApA
500 µL		H_2O

1 L 96% ethanol:1 M NH_4OAc, pH 5.5 (1:1, v/v)
100 mL enzyme buffer:50 mM Tris-HCl, pH 8.0, 5 mM $MgCl_2$ and 0.5 mM EDTA.

Procedure

A mixture of 20 µL PNPase (5 mg/mL), 3 µL 5 M NaCl, 16 µL water and 1 µL trypsin (1 mg/mL) is incubated at 37°C. After 0, 15, 30, 45, 60, 90, and 120 min 5 µL aliquots are added to a solution of 1 µL trypsin inhibitor in 19 µL enzyme buffer each, and stored in ice. When the trypsin in the final sample has been inactivated, all samples are tested for their activity in primer dependent and independent polymerisation by mixing 5 µL of PNPase solution with 10 µL test mixture (-primer) and (+primer) on a test plate and incubating it for 60 min at 37°C. Each experiment is done in duplicate according to the following scheme for every solution.

polymerisation mix (-primer) µL	polymerisation mix (+primer) µL	enzyme solution µL	
10	-	5	
10	-	5	
-	10	5	t = 0
-	10	5	
10	-	5	
10	-	5	
-	10	5	t = 15
-	10	5	

The mixture from the test plates are transferred to 1 cm x 11 cm paper strips applying the sample about 1 to 2 cm from the upper edge of the strip. The strips are dried with a fan and are chromatographed in ethanol:NH_4OAc until the solvent front reaches the lower edge of the strip (45-60 min). The strips are dried under an ir-lamp. Under uv-light (256 nm) dark spots can be seen at the origin and near the front, which are marked with a pencil, cut out and counted separately in 2 mL toluene containing 2% diphenyl oxazol. The enzyme activity is measured in µmol adenosine 5'-phosphate polymerised in 1 h at 37°C and calculated using the following formula:

$$\text{units/mL} = \frac{0.48 \cdot cpm_{start} \cdot 200 \cdot 5}{cpm_{start} + cpm_{front}}$$

This formula gives the enzyme concentration after trypsin treatment. The activity with and without primer is plotted as a function of incubation time with trypsin.

For preparative purposes a mixture consisting of

400 µL	5 mg/mL	PNPase
60 µL	5 M	NaCl
320 µL		H_2O
20 µL	1 mg/mL	trypsin

is incubated at 37°C for the optimal incubation time. The proteolytic action of trypsin is stopped by the addition of 20 µL (1 mg/mL) trypsin inhibitor. This PNPase^p is used for the synthesis of oligonucleotides of defined sequence.

2.3 Preparation of trinucleoside diphosphates

The reaction conditions for the synthesis of trinucleoside diphosphates with 3'-terminal uridine, cytidine, or adenosine are given in the following section. The elongation with GDP requires special conditions (see next pages), whereas the purification procedures are identical for all oligonucleotides. Longer oligonucleotides can be obtained by decreasing primer concentration or increasing diphosphate concentration.

The following mixture is incubated in a glass tube at 37°C for 72 h:

volume	concentration	substance
450 µL	1 M	Tris-HCl, pH 9.5
30 µL	1 M	$MgCl_2$
90 µL	5 M	NaCl
80 µL		H_2O
1 000 µL	50 mM	dinucleoside phosphate
150 µL	160 mM	nucleoside diphosphate
1 200 µL	150 U/mL	PNPase^p

2.3.1 Synthesis of A-U-G with the addition of RNase T_1 [9]

The following mixture is incubated in a glass tube at 37°C for 24 h:

volume	concentration	substance
150 µL	50 mM	ApU[NH_4^+]
	16 mg	GDP[K^+]
10 µL	1 M	$MgCl_2$
200 µL	1 M	Tris-HCl, pH 8.5
1 µL	5 x 10^5 U/mL	RNase T_1
539 µL		H_2O
400 µL	150 U/mL	PNPase^p

All reactions are terminated by heating to 80°C for 5 min. After cooling to 37°C, 25 µL alkaline phosphatase (400 U/mL) per mL reaction mixture is added and incubation is continued at 37°C for 2 h. The enzymes are partially inactivated by heating to 80°C for 10 min.

Purification of oligonucleotides:

Following incubation with phosphatase the solution is diluted to lower the salt concentration to 50 mM and applied to a DEAE cellulose column (2 cm x 25 cm) at a flow rate of 1 $mL \cdot min^{-1}$. The oligonucleotides are eluted with a linear gradient of 1 000 mL + 1 ooo mL from 10 mM $(Et)_3NH \cdot HCO_3$ to 500 mM, the flow rate is 1 $mL \cdot min^{-1}$, and fractions of 10 mL are collected. Complete separation is achieved up to the octamer, oligonucleotides with longer chain lengths may be separated on a QAE Sephadex column.

The fractions containing the trinucleoside diphosphate and longer oligonucleotides are pooled in a 250 mL round bottom flask and volume and absorption at 260 nm and 280 nm are determined to calculate the yield. The solution is evaporated to dryness with a rotary evaporator (bath temperature 40°C). Residual $(Et)_3NH \cdot HCO_3$ is removed by threefold evaporation with methanol:water (1:1, v/v). The residue is dissolved in 5 mL water and passed through a column of 10 mL Dowex 50[NH_4^+]. The column is eluted with additional 10 mL water, the effluent is sampled in a 25 mL flask with long tapering neck, concentrated to 1 mL, and further purified by passage over a Sephadex G-25 column (1 cm x 50 cm, flow rate 40 mL/h). The oligonucleotide is eluted in the void volume (13-15 mL). It is concentrated to either 100 A_{260} units/mL or is lyophylised.

Oligonucleotides which have been prepared with the addition of ribonuclease T1, have to be purified by preparative paper chromatography using an alkaline solvent. 400 A_{260} units of A-U-G may be applied to a 58 cm x 60 cm sheet of Schleicher and Schüll 2316 paper. The chromatogram is developed in isopropanol:H_2O:2 M NH_4OH (7:2:1, v/v). Although tedious this procedure up to now is the only method to guarantee an A-U-G preparation absolutely free of ribonuclease T1.

2.4 Synthesis of the block oligonucleotide A-U-G-$(U)_n$

Procedure

The following reaction mixture is incubated for 4 h at 37°C:

volume	concentration	substance
	3 µmol (100 A_{260} units)	A-U-G
	18 mg	UDP
300 µL	1 M	Tris-HCl, pH 9.5
240 µL	5 M	NaCl
10 µL	1 M	$MgCl_2$
800 µL		H_2O
600 µL	150 U/mL	PNPase^p

The reaction is terminated by heating the mixture to 80°C for 5 min. After cooling to 37°C 50 µL (400 U/mL) alkaline phosphatase is added and the solution is incubated at 37°C for 2 h (to dephosphorylate UDP). Following incubation the mixture is diluted to 12 mL and applied to a QAE Sephadex column (1 cm x 50 cm, equilibrated in 50 mM Tris-HCl, pH 7.5 and 100 mM NaCl, flow rate 40 mL/h maintained at 50°C). The column is washed with 100 mL of the same solution, and the separation of the oligonucleotides is carried out at 50°C with a linear gradient of NaCl in 50 mM Tris-HCl, pH 7.5 from 100 mM NaCl to 700 mM NaCl in a total volume of 2 000 mL (flow rate 60 mL/h). Fractions of 5 mL are collected.

The resulting oligonucleotide fractions are pooled and concentrated with a rotary evaporator until NaCl precipitates. The precipitate is dissolved by adding a small amount of water, and the oligonucleotide is desalted by passage over a Biogel P2 (2 cm x 50 cm, flow rate 80 mL/h). The oligonucleotide is eluted in the void volume, the volume applied for desalting should not exceed 40 mL (∿ 1/4 of the column volume). The oligonucleotide is concentrated with a rotary evaporator, adjusted to 20 A_{260} units/mL and stored frozen at -20°C. The chain lengths of the oligonucleotides can be determined by hplc. The base composition is determined as described by Fritz in this volume.

2.5 Preparation of polynucleotides with primer independent polynucleotide phosphorylase

Polynucleotide phosphorylase is a valuable tool for the preparation of polynucleotides. However, the reaction conditions have to be worked out for each preparation in question. If copolymers from two or more nucleoside diphosphates are synthesised, there exist always a discrepancy between the ratio of nucleoside diphosphates used in the reaction mixture and the ratio which is found in the prepared polymer [17].

Procedure

The following "standard mix" for the polymerisation of ADP, CDP, and UDP is prepared.

350 µL	1 M	Tris-HCl, pH 9.5
25 µL	1 M	$MgCl_2$
25 µL	40 mM	EDTA-K
1 200 µL	100 mM	NDP
900 µL		H_2O

Before a preparative run is started, the activity of the PNPase and the quality of the mix is tested on an analytical scale. 10 µL of the "standard mix" together with 5 µL of enzyme solution is applied to a microliter plate which is incubated at 37°C for 1 h. The mixture is applied to a 1 cm x 12 cm paper strip 1-2 cm from the upper edge. Chromatographic separation is done in ethanol:1 M NH_4OAc, pH 5.5 for 30-45 min.

The paper strip is dried under an ir-lamp and inspected under an uv-lamp. The polymer stays at the origin whereas the nucleoside diphosphate is found near the front. The yield should be 10-20% as visualised.

The polymerisation reaction is carried out in a preparative scale when the yield is in the expected range.

200 µL "standard mix" are incubated at 37°C in a 10 mL screw cap tube with 100 µL enzyme. Following the procedure outlined above, the polymerisation is controlled by running analytical tests at 1 h intervals. When the amount of polymer formed remains constant, the reaction is stopped by the addition of 500 µL H_2O and 1 mL phenol saturated with H_2O. Phenol extraction is done by shaking vigorously on a vortexer for 1 min. The layers are separated by centrifugation at 5 000 rpm for 5 min. Phenol extraction is repeated three times. The phenol layers are combined and reextracted with 0.5 mL H_2O. The combined aqueous phases are extracted 5 times with ether to remove traces of phenol. The aqueous phase is then applied to a Sephadex G-50 column (1 cm x 50 cm) which is eluted with water. Elution is monitored with an continous flow spectrophotometer. The polynucleotide eluted in the void volume (13-15 mL) is collected into a 25 mL flask and concentrated to dryness with a "rotavapor". It is dissolved in water to give a final concentration of 20 A_{260} units/mL.

2.5.1 Synthesis of polyguanylic acid and copolymers from ppG and ppU

Polymerisation mix for ppG

250 µL	1 M	Tris-HCl, pH 8.5
5 µL	1 M	$MnSO_4$
5 µL	40 mM	EDTA-K
250 µL	100 mM	GDP
1 990 µL		H_2O

The analytical test is carried out as described before, except that the ratio, mix to enzyme, is increased to 1:2 and the incubation temperature is raised to 50°C. The same changes are made in the preparative run. A longer incubation time can be used, since with poly(G) no phosphorolysis occurs.

Polymerisation mix for the synthesis of poly(U,G)

350 µL	1 M	Tris-HCl, pH 9.5
25 µL	1 M	$MgCl_2$
25 µL	40 mM	EDTA-K
200 µL	100 mM	UDP
1 000 µL	100 mM	GDP
900 µL		H_2O

Equal volumes of mix and enzyme are used for the polymerisation. Incubation is done at 37°C. The ratio of uridine to guanosine in the polynucleotide should be about one. It has to be controlled by base analysis on hplc.

REFERENCES

[1] Singer, M.F. (1966) in Procedures in Nucleic Acid Research, Vol. 1, pp. 245, Harper, New York, N.Y.

[2] Klee, C.B. (1971) in Procedures in Nucleic Acid Research, Vol. 2, pp. 896, Harper & Row, New York, N.Y.

[3] Schetters, H., Gassen, H.G., and Matthaei, H. (1972) Biochim. Biophys. Acta 272, 549

[4] Kierkegaard, L.H. (1973) Biochemistry 12, 3627

[5] Letendre, C.H. and Singer, M.F. (1974) J. Biol. Chem. 249, 7383

[6] Thach, R.E. (1966) in Procedures in Nucleic Acid Research, Vol. 1, pp. 520, Harper, New York, N.Y.

[7] Thach, R.E. and Doty, P. (1965) Science 147, 1310

[8] Thach, R.E. and Doty, P. (1965) Science 148, 632

[9] Mohr, S.C. and Thach, R.E. (1969) J. Biol. Chem. 244, 6566

[10] Thach, R.E., Sundararajan, T.A., Dewey, K.F., Brown, J.C., and Doty, P. (1966) Cold Spring Harbor Symp. Quant. Biol. 31, 85

[11] Martin, F.H., Uhlenbeck, O.C., and Doty, P. (1971) J. Mol. Biol. 57, 201

[12] Küppers, B. and Sumper, M. (1975) Proc. Natl. Acad. Sci. USA 72, 2640

[13] Walker, G.C. and Uhlenbeck, O.C. (1975) Biochemistry 14, 817

[14] Wagner, R. (1973) Diplomthesis, Münster

[15] Sninsky, J.J., Bennett, G.N., and Gilham, P.T. (1974) Nucl. Acids Res. 1, 1665

[16] Sundararajan, T.A. and Thach, R.E. (1966) J. Mol. Biol. 19, 74

[17] Brenneman, F.N. and Singer, M.F. (1964) J. Biol. Chem. 239, 893

ENZYMATIC SYNTHESIS OF ^{32}P-OLIGONUCLEOTIDES

Olke C. Uhlenbeck

Department of Biochemistry
University of Illinois

SUMMARY

The experiment describes a RNA-RNA coupling. [γ-^{32}P]ATP is prepared in the first stage and used in the kinase reaction for preparation of the donor. After the coupling reaction with T4 RNA ligase the product has an internal ^{32}P-label. The reaction products are analyzed by polyacrylamide gel electrophoresis and the oligomers eluted and recovered by ethanol precipitation (see in addition Ohtsuka and Eckert).

1 INTRODUCTION

T4 RNA ligase (EC 6.5.1.3) is an efficient enzymatic coupling reagent for oligonucleotides [1,2]. It catalyzes the following reaction:

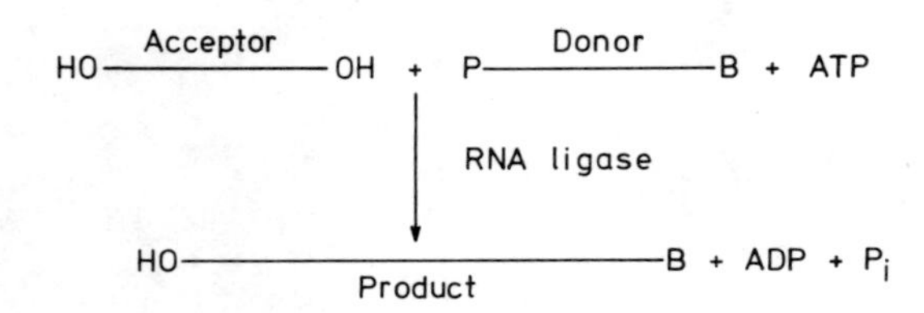

The acceptor oligomer can be any oligomer three nucleotides or longer. Ribonucleotides are much better acceptors than deoxyribonucleotides. The donor oligomer can be a nucleoside 3',5'-biphosphate or longer. Oligoribonucleotides, oligodeoxyribonucleotides, and a wide variety of modified nucleoside 3',5'-biphosphates are effective donors. The 3'-blocking group (B) is required to prevent multiple additions of the donor to acceptor. Since RNA ligase acts on unprotected single-stranded oligonucleotides, fragments of natural RNA molecules or oligonucleotides prepared with polynucleotide-phosphorylase can be used in semi--synthetic schemes. This allows one to focus synthetic effort on the region of interest for structure-function studies.

The 5'-phosphate on the donor molecule is introduced by the action of T4 polynucleotide kinase (EC 2.7.1.78):

$$\text{HO}\text{———}\text{B} + \text{ATP} \xrightarrow{\text{T4 kinase}} \text{P}\text{———}\text{B} + \text{ADP}$$

When the blocking group B is a phosphate, polynucleotide kinase from the T4 pseT 1 mutant must be used to avoid dephosphorylation by the endogenous 3'-phosphatase activity of T4 kinase. The isolation of the kinase is outlined in [3,4] the preparation of the ligase in [5].

Many applications of synthetic oligoribonucleotides do not require a large amount of material, but only sufficient ^{32}P-label to measure binding. In this experiment we will prepare an internally ^{32}P-labeled oligoribonucleotide and purify it on an acrylamide gel. All the enzyme reactions converting ^{32}P phosphate to oligonucleotide will be carried out sequentially in the same reaction mixture. This emphasizes the ease and speed of enzymatic synthesis procedures.

2 EXPERIMENTAL SECTION

2.1 Materials

Enzymes: 100 units pseT1 polynucleotide kinase available from NEN
200 units RNA ligase (NEN or PL Biochemicals)
Five enzymes for [γ-^{32}P]ATP synthesis supplied by Boehringer Mannheim: glycerophosphate dehydrogenase (EC 1.1.1.8, 180 U/mg, 2 mg/mL); triosephosphate isomerase (EC 5.3.1.1, 5000 U/mg, 2 mg/mL); glyceraldehyde 3-phosphate--dehydrogenase (EC 1.2.1.12, 80 U/mg, 10 mg/mL); 3-phosphoglycerate kinase (EC 2.7.2.3, 450 U/mg, 10 mg/mL); lactate dehydrogenase (EC 1.1.1.27, 550 U/mg, 5 mg/mL)

Radiochemicals:
10 mCi ^{32}P orthophosphoric acid in
1 ml of HCl-free aqueous solution (NEN, NEX 053 or Amersham)

Buffers: 20 mL E buffer:

5 mM DTT (dithiothreitol)
50 mM HEPES, pH 8.2 stored frozen (N-2-hydroxyethylpiperazine-N'-2-ethanesulfonic acid)

20 mL A mix:

250 mM HEPES, pH 8.2
60 mM $MgCl_2$
30 mM DTT
0.60 mM L-α-glycerol phosphate
0.25 mM ADP

2 mL 5X buffer:

250 mM HEPES, pH 8.3
100 mM $MgCl_2$
15 mM DTT

Materials for gel electrophoresis:

20% Urea gel stock: 232 g acrylamide, 8 g bis-acrylamide, 504 g urea, H_2O to 1 L. Heat at 37°C until in solution (2-3 h), add 15-20 g Amberlite MB3. Stir 1 h, filter and add H_2O to 1080 mL. Add 120 mL of 500 mM tris-borate, pH 8.3, 10 mM EDTA.

Additional materials:

sodium pyruvate; NAD^+; TEMED (N,N,N',N'-tetramethylethylene diamin); ammonium persulfate; ammonium sulfate, urea; PEI cellulose tlc plates (10 cm x 20 cm) Schleicher & Schüll F1440; X-ray films e.g. DuPont Cronex LO-doseplus; casettes or filmholders; disposable plastic gloves, Eppendorf microfuge tubes 1.5 mL and 0.75 mL; pipette tips; 100 mL 1 M CH_3COOK (sterile); 10 mL 1 M $MgCl_2$

Equipment:

Eppendorf microcentrifuge; adjustable pipettes o to 20 µL (several); Geiger counter; 10°C, 37°C water bath (small); radiation shields (several); lead containers for samples; radiation disposal facilities; Sorvall centrifuge (or equivalent) and glasstubes (∿ 8 mL); slab gel apparatus for each student (20 cm long x 15 cm wide by 1 mm thick); power supplies to 1000 V; X-ray film developing facility; shaking table or tumbler in cold room for gel elution.

2.2 Preparation of [γ-^{32}P]ATP

This procedure is adapted from Johnson and Walseth [6].

2.2.1 Enzymes

Mixed as $(NH_4)_2SO_4$ slurries mixture, stored at 4°C (DO NOT FREEZE) in the following proportion:

35 µL glycerophosphate dehydrogenase
1 µL triosephosphate isomerase
20 µL glyceraldehyde 3-phosphate-dehydrogenase
2 µL 3-phosphoglycerate kinase
20 µL lactate dehydrogenase

---

78 µL

Put 10 µL of mixture in 750 µL centrifuge tube.

Just prior to use, spin 5 min in Eppendorf centrifuge, remove supernatant and resuspend pellet in 140 µL of E buffer.

2.2.2 Reagent Mix

Combine 20 µL A mix (stored frozen)
2.5 µL 20 mM NAD^+ (stored frozen in H_2O)
2.5 µL 40 mM sodium pyruvate (22 mg/5 mL H_2O) made up fresh

25 µL R mix

NAD^+ and pyruvate must be added just before use to avoid breakdown.

2.2.3 Reaction

Volumes given for 10 µL reaction mixture, but can be scaled up depending on concentration of P_i.

up to 2 mCi P_i in H_2O	:	6.5 µL
R mix	:	2.5 µL
Enzymes	:	1.0 µL (add last)
		10.0 µL

incubate at 25°C for 30 min.
Remove a tiny aliquot (1 µL). Spot on a 10 cm PEI cellulose plate. Develop in 0.8 M $(NH_4)_2SO_4$ (15 min). Dry and autoradiograph (1 min). ATP (R_F = 0.3) separates well from P_i (R_F = 0.9). Reaction yield should be more than 90 percent.

When yield is ascertained to be high, heat reaction 3 min at 90°C. [γ-^{32}P]ATP should keep for several weeks if stored in 50 percent ethanol (but not always). Specific activity is about 3000 Ci/nmol or 330 pmol/mCi.

2.3 Polynucleotide kinase reaction

2.3.1 Reaction

5 µL		[γ-^{32}P]ATP, 330 pmol/mCi
5 µL	100 µM	ATP
9 µL	70-500 µM	donor oligomer$^+$
4 µL		5X
2 µL	1500 units·mL^{-1}	polynucleotide kinase (pseT1)
25 µL		total reaction volume

$^+$ The following donor oligonucleotides were used: A_6Cp; A_{13}; $C_{11}Gp$; A_6C; A_9Ap

Add enzyme last, incubate at 37°C for 120 min, then heat to 90°C for 2 min. Analysis is by PEI tlc as described in the previous section. Be sure to spot ATP standard. The phosphorylated oligomer will also be analysed by gel electrophoresis as described in the next section.

2.4 RNA ligase reaction

2.4.1 Reaction

5 µL	0.12 nmol	kinase reaction product
10 µL	0.5 nmol	acceptor *
5 µL		5X
4 µL		H_2O
1 µL	100-200 units·mL^{-1}	RNA ligase
25 µL		*acceptors used: A_6C; A_3C; A_{13}

Incubate at 16°C for at least 4 h or overnight.

2.4.2 Analysis

Pour a 20% urea slab gel (1 mm thick, at least 6 wells). First mix 35 mL gel stock with 200 µL 10% ammonium persulfate, heat to 60°C and degas for 10 min. Then add 20 µL 10% TEMED. Pour quickly, insert comb. Let stand at least 1 h or overnight.

20% urea gel stock solution: To 232 g acrylamide, 8 g bis-acrylamide and 504 g urea H_2O is added to a final volume of 1 L. Heat at 37°C until a clear solution is obtained (2-3 h), add 15-20 g Amberlite MB3. Stir 1 h, filter through a paper filter and add H_2O to 1080 mL. Add 120 mL of 500 mM Tris-borate, pH 8.3 and 10 mM EDTA.

Pre-electrophoresis is for 1 h at 1000 V. Running buffer is 50 mM Tris-borate, pH 8.3; 1 mM EDTA. Apply samples by mixing with equal volume of 10 M urea. Include well with dye markers (33% xylene cyanol and 67% bromophenol blue) in 8 M urea. Electrophoresis is at 700 V and 40-50 mA until bromophenol blue is at the bottom (∿2 h).

Turn off power, remove buffer (Caution-lower buffer vessel contains radioactivity) separate gel plates, wrap gel on one plate with plastic wrap. For autoradiography be sure to align film carefully with gel. Figure 1 shows an example for the page separation of ligated oligonucleotides.

slot No 11 12 13 14 15 16 17 18

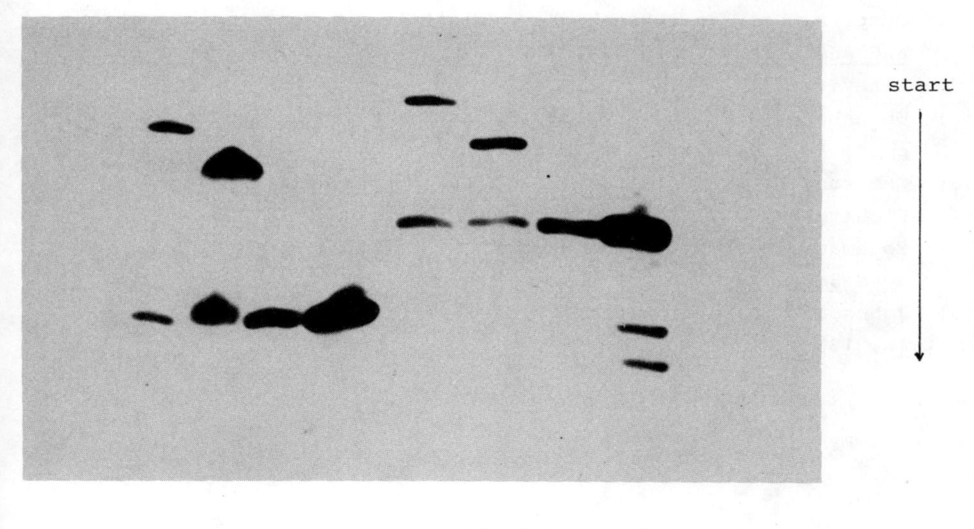

11	$*pA_6Cp + A_6C$	15	$*pA_{10}p + A_6C$
12	$*pA_6Cp + A_3C$	16	$*pA_{10}p + A_3C$
13	$*pA_6Cp + A_{13}$	17	$*pA_{10}p + A_{13}$
14	$*pA_6Cp$	18	$*pA_{10}p$

Figure 1 Page separation of ligated oligonucleotides

2.4.3 Recovery of oligomer

Align film with gel (film underneath glass). Cut out oligomer products. Put gel slice in 2 mL of sterile 1.0 M potassium acetate, pH 4.5 and shake gently 2 h or overnight at 4°C. If oligomer is large (chain length >15), it helps to crush the gel slice. (Other things people put in elution buffers include 0.1% SDS, 0.1% phenol, 10 µg/mL RNA carrier, 1 mM EDTA or anything else that they think will inhibit nucleases.)

After elution, spin out gel fragments and isolate supernatant. Add 3 volumes ethanol to the supernatant. If no carrier is present, make the supernatant 1 mM $MgCl_2$ before adding ethanol. Let tube stand at -60°C (dry ice-ethanol) for 30 min. Spin 20,000 g for 30 min at -10°C.

REFERENCES

[1] Gumport, R.I. and Uhlenbeck, O.C. (1981) in Gene Amplification and Analysis V II, p 314, J.G. Chirikjian and T.S. Papar eds., Elsevier

[2] Uhlenbeck, O.C. and Gumport, R.I. (1982) in The Enzymes Vol 15, p 31, P. Boyer ed., Academic Press

[3] Cameron, U. and Uhlenbeck, O.C. (1977) Biochemistry 16, 5

[4] Richardson, C.C. (1972) Prog. Nucl. Acid Res. 2, 815

[5] Moseman-McCoy, M.I. and Gumport, R.I. (1979) Biophys. Biochim. Acta 562, 149

[6] Johnson, R.A. and Walseth, T.F. (1979) Adv. Cyclic Nucl. Res. 10, 135

JOINING OF OLIGORIBONUCLEOTIDES BY RNA-LIGASE

Eiko Ohtsuka and Volker Eckert*
Faculty of Pharmaceutical Sciences
Osaka University
*Institut für Organische Chemie und Biochemie
Technische Hochschule Darmstadt

SUMMARY

An outline for the ligation of two short oligonucleotides by RNA-ligase is given. In the first step the donor to be is phosphorylated at the 5'-hydroxyl group with RNA-kinase and [γ-^{32}P]ATP. In the second reaction the two oligonucleotides are ligated by RNA-ligase to yield a 9-mer. Furthermore a procedure is given to introduce a 3'-phosphate group by the $NaJO_4$ and lysine catalyzed elimination of the 3'-terminal nucleoside.

1 INTRODUCTION

The discovery of RNA-ligase [1] made it possible to join shorter pieces of RNA to yield larger molecules such as tRNA [2] or RNA of defined sequence to investigate protein-nucleic acid interaction [3]. In combination with mainly polynucleotide phosphorylase [4] the use of RNA-ligase overcomes the difficulties which still exist in the chemical preparation of RNA molecules of defined sequence [5]. The ligation takes place between the free 3'-OH-end of one oligoribonucleotide called the acceptor and the 5'-phosphomonoester group of a second RNA molecule called the donor, using ATP as the energy generating cofactor (see Uhlenbeck) [6-9]. 5'-Phosphorylation can be performed by treatment of the oligonucleotide to be the donor with polynucleotide kinase and ATP [10].

A standard reaction mixture contains either 25-30 mM Tris-HCl, pH 7.5-9.6 or HEPES-NaOH, pH 8.3, 2 mM spermine, 10 mM $MgCl_2$, 10 mM dithiothreitol, 10 $\mu g \cdot mL^{-1}$ bovine serum albumine (BSA), oligonucleotide and ATP. ATP is mostly used in a 3-10fold excess over the oligonucleotide. One unit of the kinase is defined as the amount of enzyme which phosphorylates 1 nmol of RNA at 37°C for 30 min [10]. The reaction is usually terminated by heating to 90°C for 2 min. Purification of the oligonucleotide and removal of excess ATP and ADP is carried out with standard chromatographic methods such as gel filtration or ion exchange chromatography. The oligonucleotide phosphorylated by this procedure is ready for use in the ligation.

Ligation usually is carried out in the following reaction mixture: 50 mM HEPES-NaOH, pH 8.3, 10 mM $MgCl_2$, 10 mM dithiothreitol, 10 $\mu g \cdot mL^{-1}$ bovine serum albumine, 15% dimethylsulfoxide (DMSO); donor and acceptor oligonucleotide, ATP. One unit of ligase is needed to ligate 1 nmol of 5'-phosphorylated donor to a "good" acceptor [1,6,7] at 37°C for 30 min using a 2-5fold excess of ATP over donor [9,11]. If a higher excess of ATP is used the reaction rate may be slower because the enzyme becomes adenylated. The adenylated intermediate does not participate in a new reaction cycle [12-16], since the formation of the diester bond requires the free enzyme. This may happen with poor donors such as oligodeoxynucleotides. The minimum chain length for the acceptor is a trinucleotide. Increasing the length of the acceptor has only a small influence on rates and yields suggesting that the active site of the enzyme binds a trinucleotide acceptor [12,17].

The efficiency in the ligation reaction strongly depends upon the base composition within the 3'-terminal region of the acceptor molecule as demonstrated in table 1. Oligomers containing only adenylates within the three final nucleotides are the best acceptors, those with guanylate and cytidylate are intermediate and those with uridylate are the worst. Deoxynucleotides as acceptors are in general 200fold less efficient as compared to their ribonucleotide analogues [13,14,17-20]. It has been shown, that U-A-G and A-U-G are poorer acceptors compared with A-A-G [15]. Thus a single uridine within the three terminal nucleotides of the acceptor decreases the yield in ligation.

To avoid self condensation of the donor molecule often one has to block the 3'-nucleoside of the donor molecule. Various procedures are known to block the 3'-OH-group of the donor. These are the 2',3'-ethoxymethylidene group [20], phosphorylation [21] and the addition of a single 3'-protected nucleotide to the 3'-end of the oligonucleotide [22,23]. Quite often the removal of the terminal nucleoside by $NaJO_4$ and lysine at pH 8.3 is used to create an oligonucleotide with a 3'-terminal phosphate [24]. If a wild type kinase is used, which has always phosphatase activity (see Uhlenbeck) the kinase reaction is performed as the first step followed by the ß-elimination of the terminal nucleoside.

No base specificity has been found for the donor molecule, with the limitation that oligonucleotides with a 5'-terminal purine nucleotide give a twofold higher yield compared to pyrimidines [25]. The minimum size for a donor molecule is a nucleoside-5',3'-biphosphate (pNp) [12,17,26].

It has been demonstrated that with some donor molecules such as pGAUp reverse and exchange reactions between the 3'-end of the donor and the AMP molecule may occur. In this reverse reaction RNA ligase shows a strong preference for hydrolysis of the 3'-terminal phosphodiester bonds of oligoribonucleotides which possess a 3'-phosphate as blocking group. This reaction leads to unexpected side products [27]. If such reverse reactions are regarded, other blocking groups for the 3'-terminus have to be used.

An ATP-independent reaction may be used to attach mononucleotides such as 3'-ethoxyethylidine-adenosine-5'-phosphate or base-modified nucleotides or longer oligonucleotides to the 3'-end of an acceptor [23,25,28]. This avoids the adenylation of the enzyme [23,25,28].

The adenylated donor is synthesised chemically [23] and than incubated with ligase and acceptor. Thus oligonucleotides can be elongated stepwise in the 3'-direction by adding 5',3'-biphosphates [23,29]. This procedure allows to label an oligonucleotide at the 3'-terminus with ^{32}P-phosphate [21,30]. Using a combination of methods RNA ligase proves to be an excellent tool for the synthesis of ribooligonucleotides of defined sequence. Its use is strongly recommended when modified nucleotides have to be incorporated into polynucleotides i.e. tRNA.

In the experimental section the joining of pU_4 to A_5 is outlined. Here the blocking of the 3'-end of the donor is unnecessary if an excess of acceptor is used and the donor itself is a bad acceptor [19]. If necessary, blocking is performed by ß-elimination of the 3'-ultimate nucleoside of the donor.

2 EXPERIMENTAL SECTION

2.1 Phosphorylation of the donor molecules

2.1.1 Materials and equipment

[γ-^{32}P]ATP spec. act. 10 Ci/mmol; polynucleotide kinase (EC 2.7.1.78); Tris-buffer; spermine; $MgCl_2$; dithiothreitol; bovine serum albumin (BSA); triethylammonium bicarbonate (TEAB); Sephadex G-50
water bath at 37°C, chromatography column (1.0 cm x 50 cm), fraction collector with uv-monitor.

2.1.2 Procedure for the kinase reaction

20 nmol oligoribonucleotide is dissolved in a buffer mixture consisting of 25 mM Tris-HCl, pH 9.6, 2 mM spermine, 10 mM $MgCl_2$, 10 mM dithiothreitol, 10 $\mu g \cdot mL^{-1}$ BSA, 10^6-10^7 cpm of [γ-^{32}P]ATP and 10 units of polynucleotide kinase. The total volume of 0.1 mL is incubated at 37°C for 1 h. The reaction is terminated by heating to 90°C for 2 min and applied to a column (1 cm x 50 cm) of Sephadex G-50 preequilibrated with 50 mM TEAB, pH 7.5. The product is eluted with the same buffer and monitored by uv-absorbance and radioactivity (Cerenkof counting). Product-containing fractions are combined and lyophilised. The average yield is 60-80%.

To yield an oligonucleotide with a 3'-phosphate as protecting group by ß-elimination of the 3'-terminal base, 50 nmol of the phosphorylated oligonucleotide e.g. pU_6 is lyophilised and suspended in 500 µL 0.2 M lysine, pH 8.3 and 150 µL 0.5 M $NaJO_4$. This mixture is incubated at 45°C for 2 h and then applied directly to a Biogel P2-column (1 cm x 50 cm) and eluted with water. The 5',3'-bis-phosphorylated oligonucleotide is eluted in the void volume and can be used without further purification as a donor after lyophilisation.

Table 1 A comparison of acceptors in their efficiency to participate in the ligation reaction [17]

acceptor	ligase			
	14 U/mL	35 U/mL	170 U/mL	350 U/mL
ApA	0[a]	0	0	0
$(Ap)_2A$	58	84	>95	>95
$(Ap)_3A$	66	92	>95	>95
IpI		0	0	0
$(Ip)_2I$		2	12	64
$(Ip)_3I$		14	59	81
UpU		0	0	0
$(Up)_2U$		0	6	15
$(Up)_3U$		0	5	13
UpUpC		0	6	11
$UpU(pC)_3$		16	42	72
ApUpG		2	23	45
UpApG		4	14	80
UpCpG		8	46	79
UpCpA		8	18	32
GpApC		30	62	95
GpApA		60	77	88
ApApG		31	69	100

[a] Yield is expressed as fraction of acceptor converted into product.

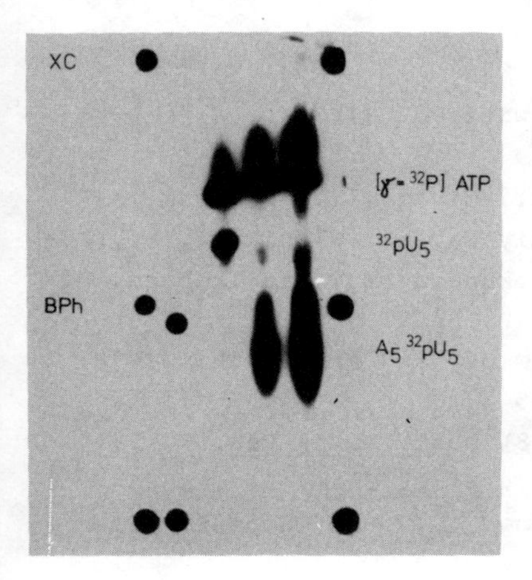

Figure 1 Separation of $[^{32}P]$-labelled oligonucleotides from a ligation experiment on DEAE cellulose tlc using homomix III
Markers: XC = Xylene cyanol
BPh = Bromophenol blue

2.2 Ligation of $[^{32}P]pU_4$ with A_5

2.2.1 Materials and equipment

Donor pU_4; acceptor A_5; ATP; $MgCl_2$; dithiothreitol; bovine serum albumin (BSA); N-2-hydroxyethylpiperazine-N-2-ethanesulfonic acid (HEPES); dimethylsulfoxide; RNA ligase; TEAB;

DEAE-Sephadex A-25; homomix III [31]; DEAE-cellulose plates (Polygram CEL 300, DEAE/HR2/15, Macherey and Nagel); X-ray film (DuPont Cronex, type 1o-dose plus); water bath at 25°C; chromatographic column (0.5 cm x 30 cm); fraction collector with uv-monitor; tlc-tank for 20 cm x 20 cm plates; drying oven set at 70°C.

2.2.2 Procedure

10 nmol donor (pU_5), 20 nmol acceptor (A_5) and 20 nmol ATP are dissolved with the buffer mixture (10 µL 0.5 M HEPES-NaOH, pH 8.3; 10 µL 0.1 M $MgCl_2$; 10 µL 0.1 M dithiothreitol 10 µL; 100 $\mu g \cdot mL^{-1}$ BSA and 15 µL DMSO to yield a total volume 0.1 mL. 10 units of RNA ligase (EC 6.5.1. 3) are added and the reaction mixture is incubated at 25°C for 2 h. The mixture is heated to 90°C for 2 min and an aliquot (ca 10^4 cpm) is analysed by homochromatography using homomix III (Fig. 1). The product can be isolated by either gel filtration on a Sephadex G-50 column (1 cm x 50 cm) or by chromatography on a column (1 cm x 20 cm) of DEAE Sephadex A-25 with a gradient of TEAB 0.05 - 1 M, total volume 150 mL. Small amounts of material are preferentially separated by polyacrylamide gel electrophoresis (see Gait and Uhlenbeck). Fractions containing the product are combined and concentrated *in vacuo*. Finally the residue is dissolved in water and lyophilised. Yield of 9-mer 20-30%.

REFERENCES

[1] Silber, R., Malathi, V.G., and Hurwitz, J. (1972) *Proc. Natl. Acad. Sci. USA* *69*, 3009

[2] Ohtsuka, E., Tanaka, S., Tanaka, T., Miyake, T., Markham, A.F., Nakagawa, E., Wakabayashi, T., Taniyama, Y., Nishikawa, S., Fukumoto, R., Uemura, H., Doi, T., Tokunaga, T., and Ikehara, M. (1981) *Proc. Natl. Acad. Sci. USA* *78*, 5493

[3] Meyhack, B., Pace, B., Uhlenbeck, O.C., and Pace, N.R. (1978) *Proc. Natl. Acad. Sci. USA* *75*, 3045

[4] Kikuchi, Y. and Sakaguchi, K. (1978) *Nucl. Acids Res*. *5*, 591

[5] Mohr, S.C. and Thach, R.E. (1969) *J. Biol. Chem*. *244*, 6566

[6] Walker, G.C., Uhlenbeck, O.C., Bedows, E., and Gumport, R.J. (1975) *Proc. Natl. Acad. Sci. USA* *72*, 122

[7] Ohtsuka, E., Nishikawa, S., Sugiura, M., and Ikehara, M. (1976) Nucl. Acids Res. 3, 1613

[8] Ohtsuka, E., Nishikawa, S., Fukumoto, R., Tanaka, S., Markham, A.F., Ikehara, M., and Sugiura, M. (1977) Eur. J. Biochem. 81, 285

[9] Uhlenbeck, O.C. and Cameron, V. (1977) Nucl. Acids Res. 7, 85

[10] Richardson, C.C. (1965) Proc. Natl. Acad. Sci. USA 54, 158

[11] Bruce, A.G. and Uhlenbeck, O.C. (1978) Nucl. Acids Res. 5, 3665

[12] Hinton, D.M., Baez, J.A., and Gumport, R.J. (1978) Biochemistry 17, 5091

[13] Moseman McCoy, M.J. and Gumport, R.J. (1980) Biochemistry 19, 635

[14] Hinton, D.H. and Gumport, R.J. (1979) Nucl. Acids Res. 2, 453

[15] Gumport, R.J. and Uhlenbeck, O.C. (1981) in Gene Amplification and Analysis, Vol. II: Analysis of Nucleic Acid Structure by Enzymatic Methods, Chirikjian, J.G. and Papas, T.S., Eds., pp 313-345, Elsevier North Holland, Inc., New York

[16] Kaufmann, G. and Kallenbach, N.R. (1975) Nature (London) 254, 452

[17] England, T.E. and Uhlenbeck, O.C. (1978) Biochemistry 17, 2069

[18] Higgins, N.P., Geballe, A.P., and Cozarelli, N.R. (1979) Nucl. Acids Res. 6, 1013

[19] Ohtsuka, E., Takefumi, D., Uemura, H., Taniyama, Y., and Ikehara, M. (1980) Nucl. Acids Res. 8, 3909

[20] Ohtsuka, E., Nishikawa, S., Markham, A.F., Tanaka, S., Miyake, T., Wakabayashi, T., Ikehara, M., and Sugiura, M. (1978) Biochemistry 17, 4894

[21] Ohtsuka, E., Uemura, H., Doi, T., Miyake, T., Nishikawa, S., and Ikehara, M. (1979) Nucl. Acids Res. 6, 443

[22] Sninsky, J.J., Last, J.A., and Gilham, P.T. (1976) Nucl. Acids Res. 3, 3157

[23] Ohtsuka, E., Miyake, T., Nagao, K., Uemura, H., Nishikawa, S., Sugiura, M., and Ikehara, M. (1980) Nucl. Acids Res. 8, 601

[24] Neu, H.G., Heppel, L.A. (1964) J. Biol. Chem. 239, 2927

[25] Uhlenbeck, O.C. and Gumport, R.J. (1982) The Enzymes, 3rd Ed., Vol. 15, pp 31-58

[26] Kikuchi, Y., Hishinuma, F., and Sakaguchi, K. (1978) Proc. Natl. Acad. Sci. USA 75, 1270

[27] Krug, M. and Uhlenbeck, O.C., in preparation

[28] England, T.E., Gumport, R.J., and Uhlenbeck, O.C. (1977) Proc. Natl. Acad. Sci. USA 74, 4839

[29] Barrio, J.R., delCarmen, M., Barrio, G., Leonard, N.J., England, T.E., and Uhlenbeck, O.C. (1978) Biochemistry 17, 2077

[30] England, T.E. and Uhlenbeck, O.C. (1978) Nature 275, 560

[31] Jay, E., Bambara, R., Padmanabhan, R., and Wu, R. (1974) Nucl. Acids Res. 1, 331

APPLICATION OF HIGH PERFORMANCE LIQUID CHROMATOGRAPHY TO OLIGONUCLEOTIDE SEPARATION AND PURIFICATION

Larry W. McLaughlin and Jörg U. Krusche*
Max-Planck-Institut für experimentelle Medizin
Göttingen
*Dupont de Nemours (Deutschland) GmbH
Abteilung Analytische Instrumente
Bad Nauheim

SUMMARY

The use of high performance liquid chromatography in combination with microparticulate bonded phase columns results in oligonucleotide separation with a high degree of resolution. Since oligonucleotides are polyanions with lipophilic bases, both anion-exchange and reverse-phase chromatography are effective for their purification. A description of a typical hplc set up as well as of differences observed for commercially available systems is included. A section on bonded phase columns includes description of their preparation as well as suggestions for user care to increase column life. An application section describes oligonucleotide separation and purification on an analytical scale (1 A_{260} unit), a semi-preparative scale (1-50 A_{260} units) and a preparative scale (>50 A_{260} units). The examples used include oligonucleotides prepared by solid phase chemical synthesis, enzymatic synthesis and enzymatic degradation techniques.

1 INTRODUCTION

High Performance Liquid Chromatography (hplc) has superceded liquid chromatography as the method of choice for isolation, purification and analysis of oligonucleotides and nucleic acid constituents. This has been primarily a result of the development in the last decade of stable chemically bonded stationary phases. This advance in the preparation of stationary phases, coupled with the advent of microprocessors, accurate small volume pumps and sensitive detectors has resulted in a fast widely used and reproducible chromatographic technique.

The process of chromatography involves the partitioning of a solute between a stationary phase and a mobile phase. A solute, such as an oligonucleotide, introduced at the inlet of a column will migrate through the column at a rate dependent on its interaction between these two phases. With a complex mixture of solutes, an ideal chromatographic separation occurs when the migration velocity of each individual component varies to the extent that they exit the column as separate and pure compounds. Since oligonucleotides are essentially polymers containing anionic phosphate groups and lipophilic bases, two types of chromatography are potentially useful in the area of nucleic acids, namely reverse-phase chromatography and ion-exchange chromatography using anion exchangers.

Reverse-phase chromatography is a type of partition chromatography introduced by Howard and Martin in 1950 [1] which has become more popular with the development of bonded phase columns. It can be compared with traditional adsorption chromatography which is performed on a polar stationary phase such as silica gel and uses a nonpolar mobile phase. In adsorption chromatography polar solutes are tightly adsorbed to the polar stationary phase and are therefore eluted later than non--polar solutes. Polar compounds can be eluted from the stationary phase by increasing the polarity of the mobile phase. In reverse-phase chromatography, as implied, the process is reversed. The stationary phase is non-polar and the mobile phase is polar. It is now the non-polar or lipophilic interactions which determine the migration velocity along the stationary phase, and as one expects, polar solutes are now eluted earlier from the column than non-polar solutes. Elution of a particular compound in this case is now effected by reducing the polarity of the aqueous mobile phase usually by the addition of organic solvents.

The fundamental mechanism of anion-exchange chromatography is the adsorption-desorption of the anionic solute on a cationic stationary phase. The adsorption and desorption are equilibrium processes which depend both upon the nature of the eluting buffer and its concentration. For a given set of conditions, desorption of a mixture of polyanionic solutes will depend upon the net charge of each polymer. The elution of polyanions such as oligonucleotides occurs according to the number of anionic charges, generally reflecting the length of the polymer, although there often exist secondary adsorptive effects involving non-ionic functional groups.

2 hplc EQUIPMENT

High performance liquid chromatography is not significantly different from other types of liquid chromatography with the exception that column inlet pressures in excess of 100 bar (1420 psi) are often necessary to maintain reasonable flow rates. A simple hplc set up is diagrammed in figure 1. One or two eluting buffers are required depending on whether isocratic or gradient elution of the column is desired. The eluting buffers feed a pump usually containing a gradient former. Following the pump is an injection port, the column, detector and recorder.

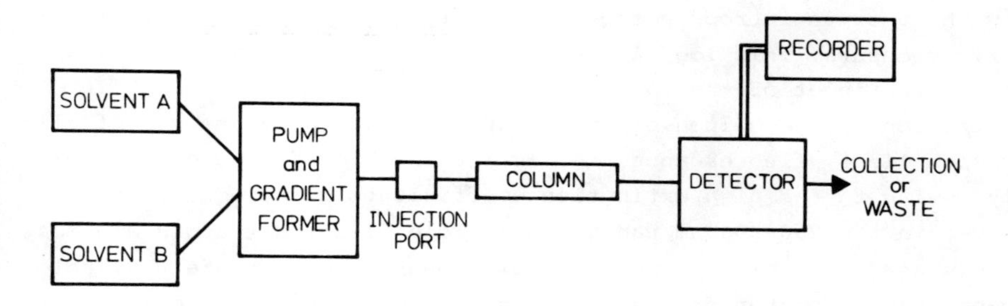

Figure 1 Components of a typical hplc system

Most of the components necessary to set up an hplc system are often already available in scientific research laboratories. However, one difficult aspect in building an hplc system is versatile and reproducible gradient formation. Commercial hplc systems employ two methods of gradient formation. In one case the gradient is mixed on the high pressure side of the pump. One pump for each buffer is required and the gradient formed by controlling the relative flow rate of each pump using a microprocessor. A high pressure mixing chamber or a simple "T" joint is used between the pumps and the column. The advantage of this system is that bubble formation as a result of solvent mixing will not affect pumping efficiency although may interfere with optical detection of the solutes. It may, however, be difficult to produce accurate gradients at the initial phase and final phase of a gradient from 100%

buffer A to 100% buffer B where either pump A or pump B must be able to deliver only microliters of solvent per min. In the second case, the gradient is mixed on the low pressure side of the pump. Two or more proportioning valves controlled by a microprocessor determine the amount of buffer from either reservoir which flows into a mixing chamber prior to reaching the pump head. The advantage of such a system is that only one pump is required, it avoids the problem of flow programming pumps and allows more accurate gradient control. With low pressure solvent mixing it is necessary to degas solvents to prevent bubble formation which will affect pumping efficiency. Both gradient systems can be employed and by carefully choosing the mixture of the two eluting buffers and the gradient to be used, the disadvantages of both systems can be minimalized.

A number of injection port designs are available which allow introduction of the sample mixture to the top of the column. Initially injection was made through a rubber septum against the back pressure of the column. While this is still a possibility, most injection systems allow the user to introduce the sample into an isolated loop at atmospheric pressure. This loop is then connected through high pressure seals to the inlet of the column.

Detection of the column effluent can be achieved by the use of ultraviolet light, fluorescence, refractive index or done electrochemically. Owing to the high extinction coefficients of oligonucleotides observed in the 250-300 nm range, ultraviolet light detectors are most commonly used. Two basic types of flowthrough detectors are available. An uv-spectrophotometer detector allows one to monitor any desired wavelength in the ultraviolet spectrum. It is also useful when working with preparative separations if detuning of the spectrophotometer is necessary as will be described later. Filter photometer uv-detectors monitor only at discrete wavelengths but are less expensive than spectrophotometers. Such a detector with a filter near 260-280 nm range is generally sufficient for most oligonucleotide separations.

3 BONDED PHASE COLUMN

The column of an hplc system, and more specifically, its stationary phase, is ultimately the important factor which determines the extent of resolution obtainable for a given oligonucleotide mixture. Chemically bonded stationary phases were first introduced by Halasz and Sebastian in 1969 [2] by treating silica gel with thionylchloride and subsequent reaction with 3-hydroxypropionitrile. In 1974 new bonding techniques were applied to silica microparticles (5-10 µm) which resulted in efficient chromatographic supports with high surface areas.

The production of bonded phase supports involves the coupling of an organic molecule to a silica particle through an ester, amine, carbon or siloxane linkage. Owing to stability and relative ease of production it is the latter variety which comprises the bulk of commercially available materials. The reaction of an organic monofunctional silane with the silica particle is controlled in such a way that monolayer coverage of the silica results. Polymeric coverage of the silica particle using trifunctional silanes is possible although undesirable since it may lead to less efficient columns. After reaction with the organic silane, many manufacturers "cap" the support using trimethylsilylchloride or similar silylating reagent. This treatment inactivates any remaining accessible Si-OH groups which otherwise would result in undesirable adsorption effects. Some Si-OH groups remain but are generally inaccessible to the solute and do not affect chromatographic resolution. Uncapped supports particularly when used in the reverse-phase mode often exhibit excessive peak tailing. Using this methodology a variety of functional groups can be bonded to silica microparticles for use in normal-phase, reverse-phase or ion-exchange chromatography. One advantage of bonded phase supports is that it allows one to adjust the polarity of the stationary phase by changing the functional group of the organic molecule.

Commercially produced prepacked columns are available and care should be taken such that a reasonable lifetime is obtained from each column. One of the simplest and easiest aspects of column care is to insure that elution buffers are filtered through membrane filters every day prior to use. This will prevent particulates from collecting on the inlet frit of the column and producing high column back pressures which often indicate the end of column life.

On the other hand, the problem of high back pressure which often occurs after a number of chromatographic separations can often be remedied by removing the column inlet frit and cleaning it in 50% aqueous nitric acid. Columns should be stored after use in either distilled water or organic solvents such as methanol. Columns, which after extensive use, exhibit severely reduced efficiency can often be significantly regenerated by washing with an organic solvent like methanol. The use of pre-columns or guard columns can also significantly increase column life. Since the irreversibly adsorbed solutes which often drastically effect column efficiency are found within the first centimeter of stationary phase, the use of a guard column will protect the chromatography column from collecting such material. Guard columns can be renewed with minimal cost.

Column cost is most effectively reduced by purchasing bulk support and learning to pack microparticulate silica columns. While reports in the literature would indicate this as very difficult process, we have found that most bonded phase supports can be packed in a methanol slurry using most any pump - such as that from an hplc system - which will produce a flow of 10 $mL \cdot min^{-1}$ and a pressure of 200 bar [3].

We have suggested that oligonucleotides are best resolved using reverse-phase or anion-exchange chromatography. However, within these two modes a number of possibilities exist as to the choice of stationary phase. Reverse-phase chromatography can be used with commercially available trimethylsilyl, ethylsilyl, octasilyl, octadecasilyl, phenylsilyl bonded phases as well as supports carrying a variety of organic functional groups (CN, NO_2, etc). Anion-exchange chromatography uses primarily tetraalkylammonium supports although commercially available amino and dimethylamino supports will also act as anion exchangers when used with acidic buffers. Majors [4] has compiled a list of hplc supports for various modes of chromatography which includes information on functional groups, particle size and manufacturer.

4 SELECTED APPLICATIONS

The applications section of this chapter is divided into two sections. The first discusses analytical and semi-preparative uses and the second describes preparative hplc. Analytical hplc resolution of oligonucleotide mixtures to determine purity or retention time is easily accomplished with about 1 A_{260} unit of a mixture of oligonucleotides or less of a pure species. Semi-preparative hplc is a rather arbitrary term, which for the following discussions will be defined as chromatography using an analytical size column (commonly 4.6 mm x 250 mm and containing about 3 g of stationary phase) where the product can be collected with one to three multiple injections. In our experience we have found that from 1 to 50 A_{260} units can be obtained in this manner. For isolation of oligonucleotide quantities in excess of 50 A_{260} units preparative hplc with large columns is often most advantageous.

4.1 Analytical and semi-preparative hplc

In order to begin hplc isolation of oligonucleotides one must decide upon the column to be used and the desired mobile phase. While either anion-exchange or reverse-phase are acceptable modes of chromatography, an argument can be made that oligonucleotide separation and purification is best accomplished using both techniques. It is possible that an oligonucleotide isolated from a complex mixture involving only one type of chromatographic interaction may have co-eluted with a similar oligonucleotide or other organic contaminant. It is less likely that an oligomer, which has been shown to be a single peak by anion-exchange and reverse-phase chromatography, contains additional by--products. The initial separation and oligomer isolation can be made using anion-exchange chromatography and then the product analyzed by reverse-phase chromatography, and if necessary re-isolated. The reverse process is also possible, although in our experience it is easier to isolate oligomers initially based on the number of phosphates or polymer length and then observe if selection by lipophilic interactions indicates more than one product.

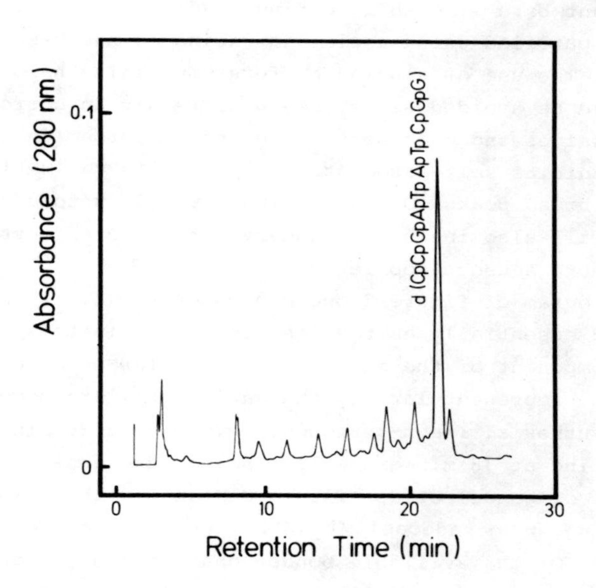

Figure 2 Chromatography of an oligonucleotide mixture containing the decamer d(CpCpGpApTpApTpCpGpG) (courtesy of Prof. M.J. Gait, Cambridge, England) on a 4.6 mm x 250 mm Whatman SAX column. Solvent A: 0.001 M KH_2PO_4, pH 6.3 in formamide:H_2O (60:40, v/v). Solvent B: 0.3 M KH_2PO_4, pH 6.3 in formamide:H_2O (60:40, v/v). Flow: 2 mL/min; temperature: 45°C; detector: 280 nm, 0.16 aufs; gradient: 0-75% solvent B, 40 min

Anion-exchange chromatography is a relatively simple process for the analysis of oligonucleotides prepared by solid phase synthesis. Generally, the undesirable side products obtained after the work-up of an oligonucleotide are the various failure sequences resulting from one or more inefficient coupling reactions. It is always the case then that the desired product is also the longest oligonucleotide. The chromatogram of figure 2, shows an analysis on a strong anion exchange column of the 10-mer d(CpCpGpApTpApTpCpGpG) synthesized according to the solid support method described by Gait [5]. The elution buffer is potassium phosphate in a mixture of formamide and water also as described by Gait. Phosphate is common as the eluting buffer for anion-exchange columns since it has a high exchange potential and competes favourable with the oligonucleotide for the cationic sites of the stationary phase. It should be noted that commercially available potassium phosphate contains high content of uv-absorbing contaminants which result in significant base line shift during gradient elution. Phosphate solutions can be purified using chelating resins as has been previously described [6]. The use of eluting buffers containing halogens or formic acid should be avoided since they will result in corrosion of stainless steel valves and pump heads. The use of formamide in the eluting buffer inhibits self-association of the oligonucleotide chains which results in broad peaks. Often heating the column to a temperature of 50-60°C will also inhibit secondary structure effects and allows the use of pure aqueous mobile phases.

In the chromatogram of figure 2 one can observe that the desired product is eluted essentially as the last peak and, in this case, is also the major component of the mixture. The smaller peaks with shorter retention time represent largely the various failure sequences generated during synthesis. The n-1 sequence eluting at 20 min and the n-2 sequence eluting at 18 min are well resolved and easily observed.

This identical oligonucleotide mixture containing the 10-mer was also analyzed using an octadecasilyl (ODS) reverse-phase column as is shown in figure 3. Of the available bonded phase supports for reverse-phase chromatography, the ODS (C-18) or octasilyl (C-8) appear to give the best resolution of oligonucleotide mixtures and are the most popular reverse-phase supports. Recently we have observed that phenyl bonded supports are also useful for oligonucleotide chromatography [7]. The eluting solvent in reverse-phase chromatography is an aqueous buffer containing acetonitrile, methanol or tetrahydrofuran. In gradient elution it is the concentration of the organic solvent component which is increased to reduce polarity of the aqueous mobile phase and thus effect migration of an oligonucleotide along the stationary phase. Buffers containing acetate or phosphate are commonly used since they seem to result in better peak shape. They are most effective in increasing hydrophobic interactions between the solute and the stationary phase and thus may enhance adsorption and reduce peak tailing [8]. A second aspect of reverse-phase chromatography, that of ion-pair partitioning will be discussed elsewhere in this book.

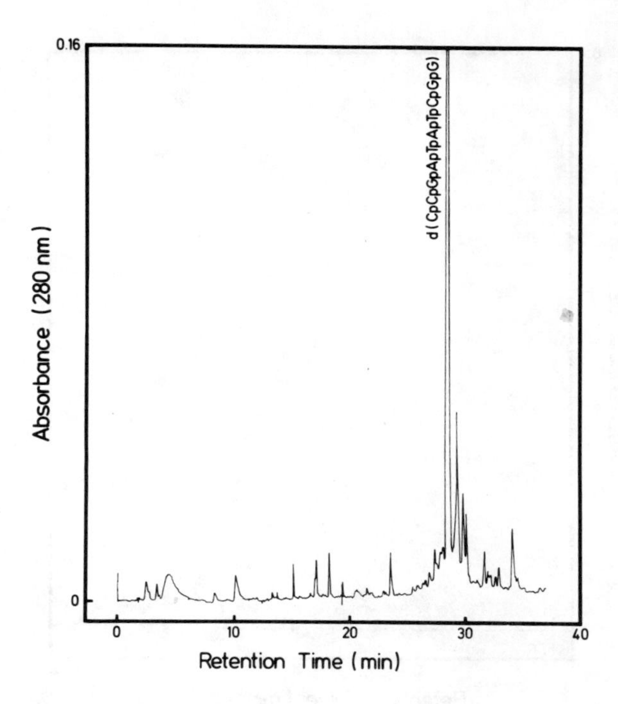

Figure 3 Chromatography of the mixture of figure 2 on a 4.6 mm x 250 mm Zorbax®-ODS column. Solvent A: 0.01 M triethylammonium acetate, pH 7.0. Solvent B: 60% acetonitrile in 0.01 M triethylammonium acetate, pH 7.0. Flow: 2 mL/min; temperature: 45°C; detector: 280 nm, 0.16 aufs; gradient: 0-30% Solvent B, 60 min

In comparing the chromatograms of figures 2 and 3, it is obvious that while in the anion-exchange mode the largest oligonucleotide elutes as the last peak in the chromatogram, the same is not observed by reverse-phase chromatography. It is still obvious which peak is the product in the chromatogram of figure 3 since it was in this case the major component, although now it elutes in the middle of the gradient. Problems can occur in the isolation of an oligonucleotide from a synthesis which was not as successful in each coupling as this one. In such a case a number of similar sized peaks representing the product, the n-1 sequence, n-2 sequence, etc. would result. By anion-exchange chromatography the product remains the longest oligomer and the last peak. However, by reverse-phase chromatography alone, it would be difficult to determine which peak was the product oligomer.

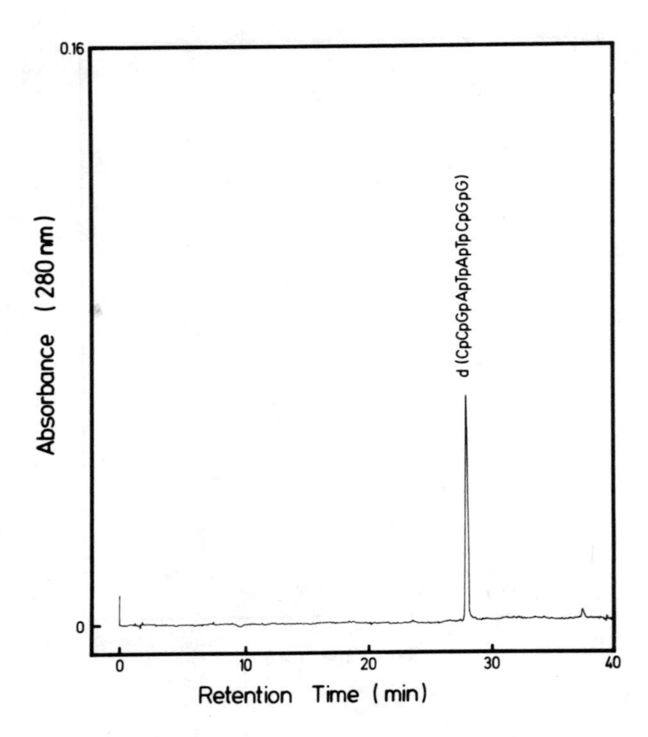

Figure 4 Chromatography of the decamer d(CpCpGpApTpApTpCpGpG), (courtesy of Prof. M.J. Gait, Cambridge, England) isolated from a Whatman SAX column on a 4.6 mm x 250 mm, Zorbax®-ODS column. Conditions as in figure 3

It is in any case recommended that when the oligonucleotide is isolated by one mode of chromatography it is subsequently chromatographed using a second mode to check purity. For example, a second larger sample of the mixture shown in figure 2 was chromatographed and the major peak eluting at 22 min was collected and desalted (Biogel P_2 or Sephadex G-10). A small aliquot of this material was then chromatographed on an ODS column with the gradient described in figure 3. The chromatogram produced is shown in figure 4. It is clear that the material isolated from the anion-exchange column co-elutes with the largest peak observed in figure 3. The chromatogram of figure 4 also indicated that the oligonucleotide is pure as judged by gradient anion-exchange and gradient reverse-phase chromatography. However, it should be noted that resolution by gradient elution is not as sensitive as in isocratic elution. Therefore, once the reverse-phase elution conditions are determined using a gradient, a second chromatogram should be run using isocratic conditions with the elution buffer containing slightly less organic solvent than indicated by the gradient chromatogram.

The anion-exchange chromatography described in figure 2 has used tetraalkylammonium salts as the cationic stationary phase. As mentioned earlier, bonded phases containing amino groups can, in acidic buffers, also function as anion exchangers. We have had considerable success using aminopropylsilyl (APS) bonded phases for anion-exchange chromatography of oligonucleotides. Therefore, for comparative purposes, the oligonucleotide mixture used in figures 2 and 3 was additionally chromatographed on an APS column with a phosphate gradient and is reproduced in figure 5. A comparison of the chromatogram of figures 2 and 5 indicate that the APS support resolves the mixture into more components than did the strong anion exchanger (SAX). Additionally the APS support produces sharper peaks with less tailing than was observed in figure 2.

With this mixture of oligonucleotides it is clear that either the APS or the SAX support result in an acceptable oligonucleotide resolution. We have observed, however, that certain advantages exist with the APS stationary phase. The amino function acts as an anion exchanger only when it is in the protonated form. The number of binding sites produced on the stationary phase and available to bind the oligonucleotide is pH-dependent. At low pH oligonucleotides are strongly adsorbed while near neutral pH a much weaker interaction occurs. This characteristic allows considerable control of the retention time for a given oligomer. Consider for example chromatography of the oligoribonucleotide $(Ap)_9A$ on an APS support. With a 60 min linear gradient from 0.05 M to 0.9 M KH_2PO_4 at pH 4.5, this decamer is eluted near the end of the gradient. If the same gradient is again used but with a phosphate buffer at pH 6.5 the retention time decreases to 10 min. Nevertheless, the resolution at pH 6.5 between $(Ap)_9A$ and $(pA)_{10}$ (containing one additional phosphate) occurs with baseline separation. It is also possible with this stationary phase to use pH gradients at constant buffer concentration for oligonucleotide separations.

While we have had success purifying oligonucleotides by a combination of anion-exchange and reverse-phase hplc it should also be noted that reverse-phase chromatography can be used exclusively in cases where the oligonucleotide of interest carries a highly lipophilic protecting group such as the monomethoxytrityl (MMT) or dimethoxytrityl (DMT) group. If during solid phase oligonucleotide synthesis a "capping" step is used after each coupling step, only the desired product will carry the MMT or DMT group as a result of the last coupling step. The oligonucleotide base and phosphate residues can then be deprotected such that the final mixture contains a number of completely de-

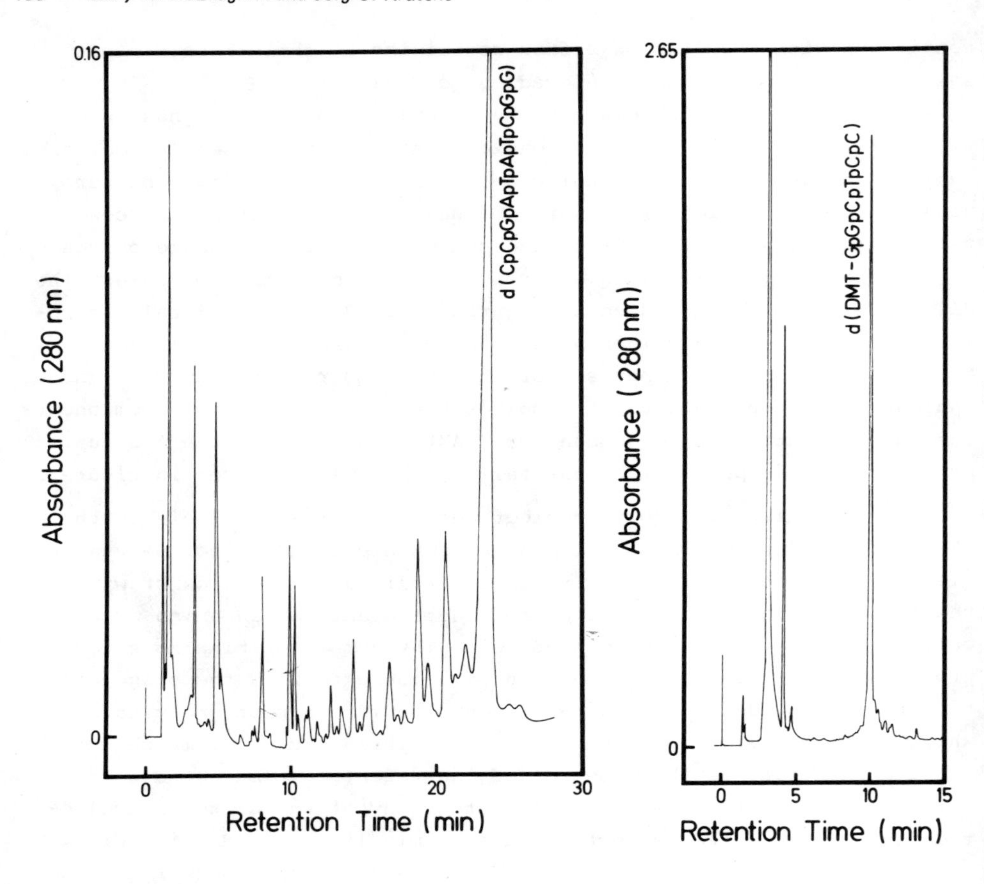

Figure 5 Chromatography of the mixture of figure 2 on a 4.6 mm x 250 mm Zorbax®-NH_2 column. Solvent A: 0.01 M KH_2PO_4, pH 4.5. Solvent B: 0.9 M KH_2PO_4, pH 4.5. Flow: 2 mL/min; temperature: 60°C; detector: 280 nm, 0.16 aufs; gradient: 0-50% solvent B, 50 min

Figure 6 Chromatography of an oligonucleotide mixture containing the hexamer d(DMT-GpGpCpTpCpC), (courtesy of H. Seliger, Ulm, F.R.G.) on a 4.6 mm x 250 mm Zorbax®-ODS column. Solvent A: 0.1 M triethylammonium acetate, pH 7.0. Solvent B: 50% CH_3CN in 0.1 M triethylammonium acetate, pH 7.0. Flow: 1.5 mL/min; temperature: 35°C; detector: 280 nm, 2.56 aufs; gradient: 50-100% solvent B, 20 min

blocked failure sequences and the partially deblocked product oligomer containing a 5' DMT or MMT group. The trityl group, being highly lipophilic, is tightly bound to an ODS stationary phase. Chromatography of this mixture will result in a chromatogram similar to that shown in figure 6 for the oligomer d(DMT-GpGpCpTpCpC) produced by Seliger and as described by Caruthers [9]. The strength of the eluting buffer can be chosen such that the failure sequences are eluted very early from the column and the DMT-oligomer is easily separated and collected. While this is a single step isolation of the desired sequence, a second isolation still is required, after removal of the DMT group, to yield the fully deblocked oligonucleotide.

In addition to chemical synthesis of oligonucleotides enzymatic techniques have proven valuable. For the preparation of RNA fragments, T4 RNA ligase has been shown to be a powerful tool with or without chemical synthesis [10,11]. Since it is often the case when using RNA li-

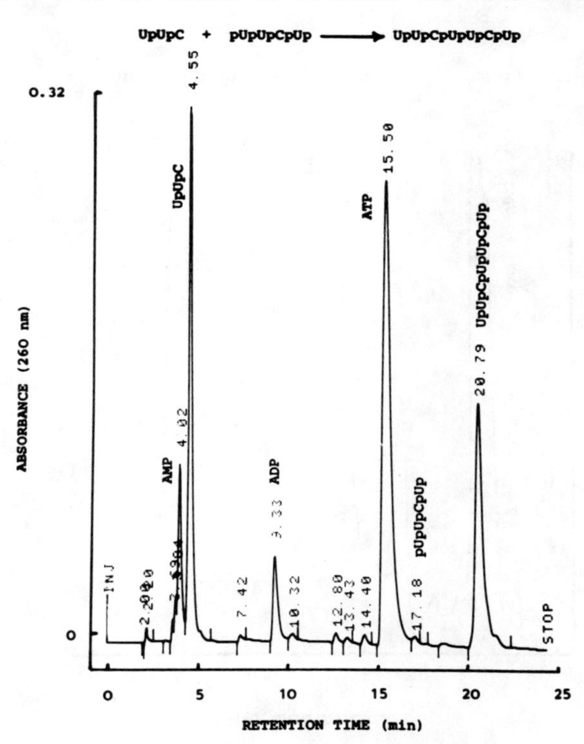

Figure 7 Analytical chromatography of an oligonucleotide mixture resulting from the ligation of pUpUpCpUp to UpUpC on a 4.6 mm x 250 mm Zorbax®-NH_2 column. Solvent A: 0.05 M KH_2PO_4, pH 4.5. Solvent B: 0.9 M KH_2PO_4, pH 4.5. Flow: 2 mL/min; temperature: 35°C; detector: 260 nm, 0.32 aufs, gradient: 0-50% solvent B, 30 min

gase that the reaction does not go to completion, a complex mixture of oligonucleotides results.

hplc is again the method of choice for the resolution and isolation of the desired oligomer. For example an analytical scale hplc separation of a mixture containing the heptamer UpUpCpUpUpCpUp prepared from an "acceptor" molecule UpUpC and a "donor" molecule pUpUpCpUp is shown in figure 7.

In this case excess acceptor oligonucleotide has been used in order to drive the reaction to completion with respect to the donor molecule. Additionally, the enzyme requires ATP which is subsequently converted to AMP. The analysis is performed by anion-exchange on an APS support such that the heptamer is the last peak eluted.

40 A_{260} units of UpUpCpUpUpCpUp was isolated semi-preparatively using a 4.6 mm x 250 mm analytical APS column and three multiple injections. One of the semi-preparative chromatograms is shown in figure 8.

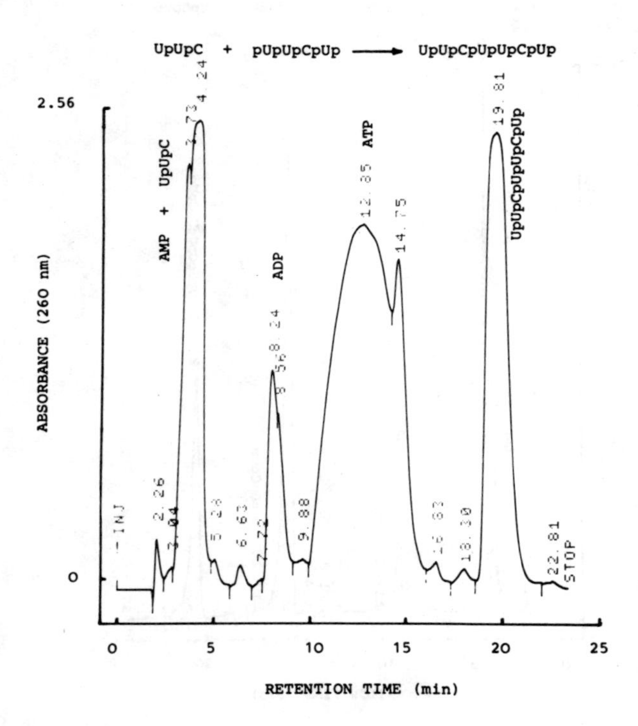

Figure 8 Semi-preparative isolation of the heptamer UpUpCpUpUpCpUp. Column and conditions as in figure 7 with detector sensitivity at 2.56 aufs

From each of the three injections approximately 15 A_{260} units of oligonucleotide was isolated. The peaks shown in figure 8 are no longer as sharp as was observed in the analytical chromatogram. This is due in part to overloading the detector capacity at its lowest sensitivity setting and also as a result of column loading effects.

After isolation and desalting of the oligonucleotide, an aliquot was analyzed by reverse-phase chromatography as is shown in figure 9. In this case the chromatogram indicated the product was 90-95% pure. A second semi-preparative isolation was then performed, using reverse--phase conditions. This resulted in an oligomer which eluted as a single peak from either column.

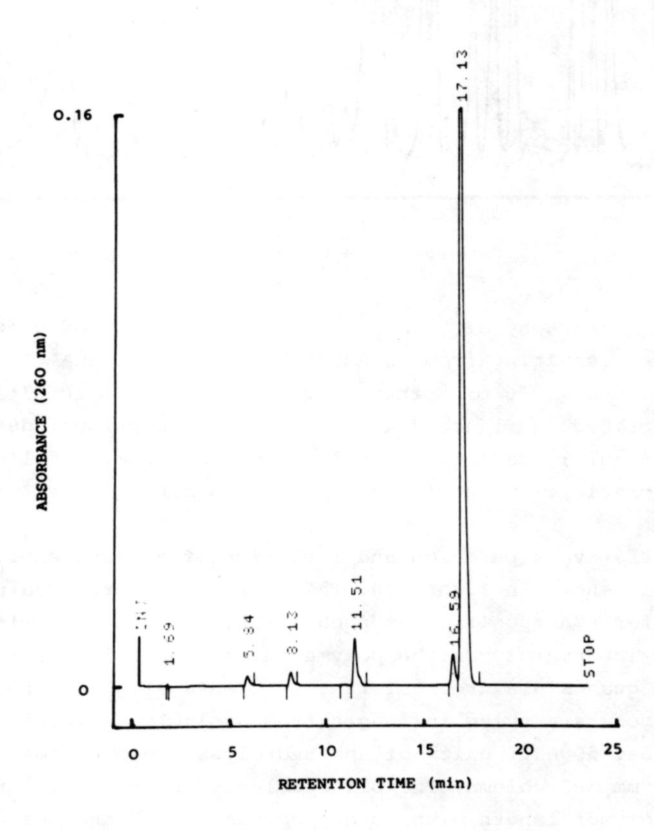

Figure 9 Re-analysis of the oligonucleotide UpUpCpUpUpCpUp isolated as in figure 8 on a 4.6 mm x 250 mm ODS-Hypersil column. Solvent A: 0.02 M KH_2PO_4, pH 5.5. Solvent B: 70% methanol in 0.02 M KH_2PO_4, pH 5.5. Flow: 2 mL/min; temperature: 35°C; detector: 260 nm, 0.16 aufs; gradient: 0-50% solvent B, 30 min

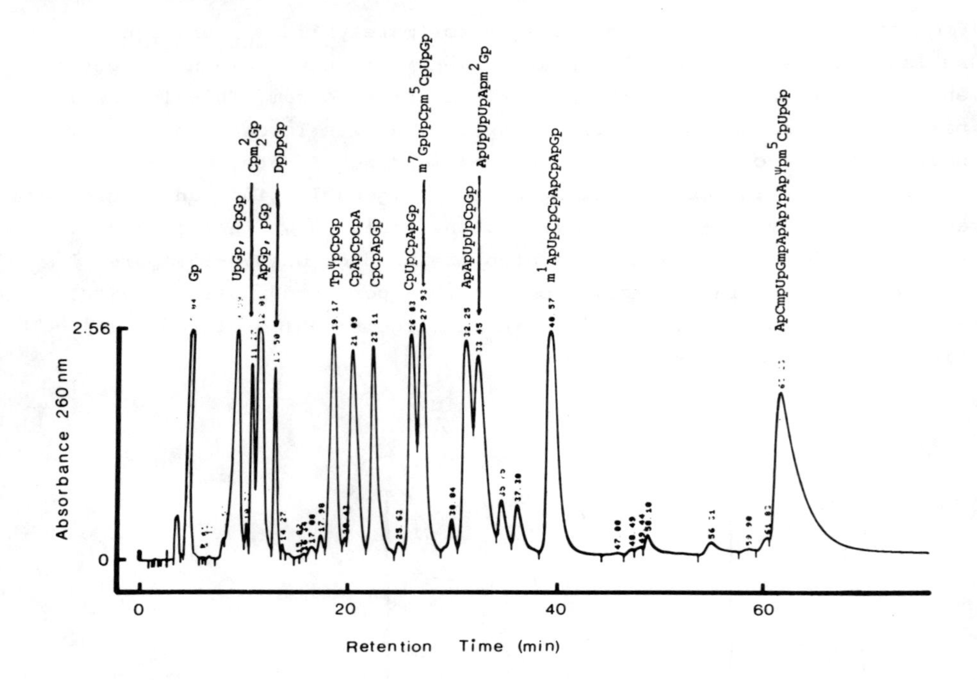

Figure 10 Chromatography of 50 A_{260} units of a mixture of oligonucleotides resulting from an RNase T_1 digest of $tRNA^{Phe}$ yeast on a 4.6 mm x 250 mm Zorbax®-NH_2 column. Column conditions as described in figure 7 with following exceptions: detector: 2.56 aufs; gradient: 0-100% solvent B, 60 min, followed by isocratic elution with solvent B, 20 min

A semi-preparative separation and isolation of a mixture of 16 oligonucleotides is shown in figure 10. This mixture is produced upon treating transfer RNA specific for phenylalanine from yeast with ribonuclease T_1, which hydrolyzes the polymer after guanosine residues (some modified guanosines are resistant to cleavage). This produces a mixture of oligomers varying in length from a single nucleotide to a dodecanucleotide. 50 A_{260} units of the hydrolysate can be resolved on a 4.6 mm x 250 mm APS column. The oligonucleotides are eluted roughly according to polymer length with Gp as the first peak and the dodecanucleotide as the last. While some of the dinucleoside diphosphates co-elute, the longer oligomers which have been of more interest in our present research are well resolved. From this chromatogram from 1-4 A_{260} units of an oligonucleotide can be isolated. Assignment of the peaks is based upon nucleoside analysis as has been described [3].

4.2 Preparative hplc

For the isolation of nucleoside building blocks or oligonucleotides in large quantities (> 50 A_{260} units) it is possible to scale up hplc separations using larger columns. There are, however, some difficulties involved particularly with peak detection. For preparative hplc often an analytical system will suffice with the use of a preparative size column. It is, however, suggested that a preparative flow cell be used to decrease the sensitivity of the detector. Without making some attempt to change the very sensitive uv-detection system used in analytical hplc the detector will register maximum absorbance during the entire chromatographic separation. While use of a preparative flow cell in the uv-detector is of some help, misleading absorbance readings may

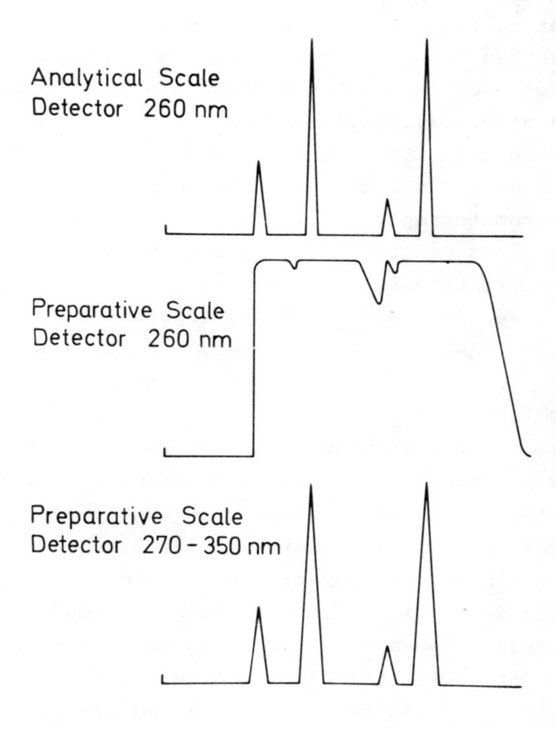

Figure 11 Stylized chromatogram illustrating the use of a preparative size hplc column for an analytical and preparative scale separation at 260 nm and a preparative scale separation with the detector detuned in the 270-350 nm range

still be observed due to the high uv-extinction coefficients observed at 260-280 nm for nucleic acid constituents. Consider for example the stylized chromatograms of figure 11. Analysis of a small aliquot of an oligonucleotide mixture on a preparative column produces a chromatogram indicating acceptable resolution similar to that shown at the top of figure 11. However, when attempting to scale up the separation 100 or 500 fold a chromatogram suggesting poor resolution and column overloading such as that shown in the center of figure 11 is produced. This "poor" chromatogram may, however, simply result from overloading the capacity of the detector. Spectrophotometric detectors can be detuned to a wavelength where the relevant extinction coefficients are now 1-10% of that observed at 260 nm. Now upon scaling up the chromatographic separation one obtains the chromatogram shown at the bottom of figure 11. It is now clear that the preparative separation occurred with resolution very similar to that observed with the analytical chromatogram. This technique can be used successfully in the isolation of the tritylated nucleosides used in oligonucleotide synthesis. The monomethoxytrityl (MMT) and dimethoxytrityl (DMT) groups are acid sensitive compounds used to protect the 5'-hydroxyl group of nucleosides during oligonucleotide synthesis. The tritylated nucleosides are usually isolated by adsorption chromatography on silica gel. However, silica gel is of itself acidic and it is sometimes the case when the resolution requires extended times, that the tritylated nucleoside is hydrolyzed during chromatography. We have observed that preparative hplc on microparticulate silica allows fast isolation of very pure tritylated nucleosides.

A chromatogram of the isolation of the MMT derivative of a protected cytidine nucleoside used in oligoribonucleotide synthesis is shown in figure 12. Normally this reaction can be driven to completion with excess monomethoxytritylchloride. We have in this case isolated the product when reaction was 90% complete in order that the elution of the starting nucleoside is visible. The isolation was performed on a 21.2 mm x 250 mm microparticulate silica column containing about 70 g of stationary phase. A preparative pump head was also employed to allow for a flow rate of 40 mL/min. With the elution conditions described in figure 12 a single injection resulted in 140 mg of the MMT--cytidine derivative. By detuning the detector to 330 nm one observes that the preparative separation looks very much like an analytical chromatogram. We are presently attempting to scale up these nucleoside isolations to a 500 mg or perhaps 1 g scale.

Preparative hplc is also useful in the isolation of milligram amounts of oligonucleotides. Consider for example the chromatogram of

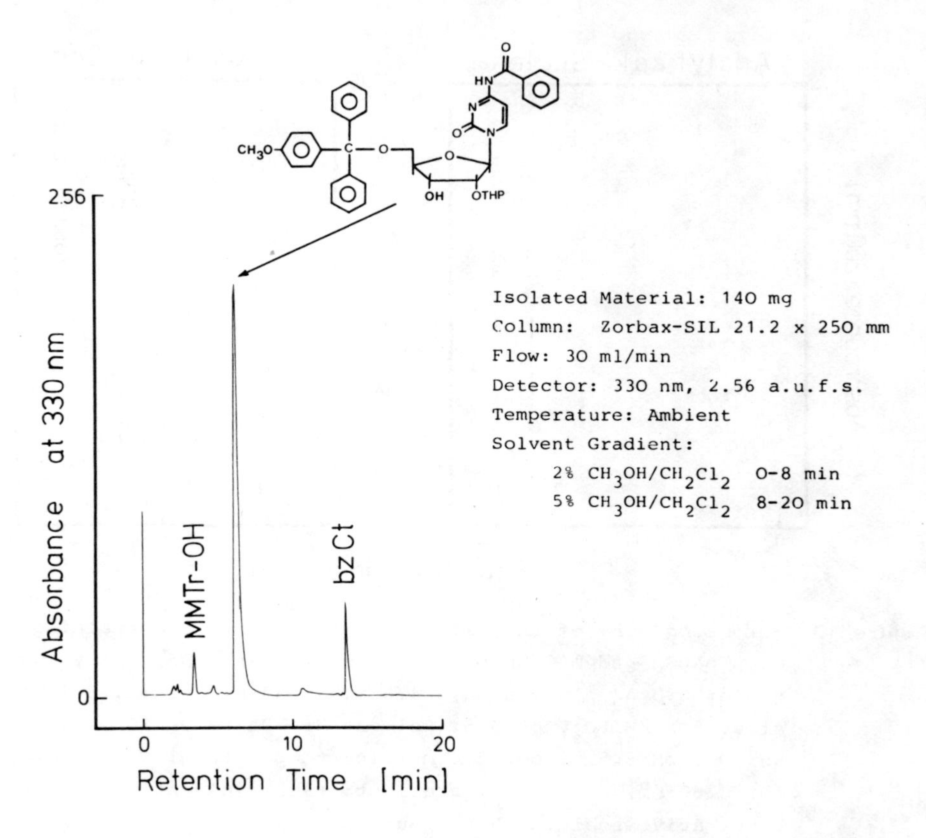

Figure 12 Preparative isolation of N^4-benzoyl-2'-O-(2-tetrahydropyranyl)-5'-O-(4-methoxytriphenylmethyl)cytidine

the ribonuclease T_1 digest of $tRNA^{Phe}$ yeast of figure 10. This digestion can be scaled up to 500 A_{260} units and separated by a single injection on a 21.2 mm x 250 mm APS column. The resolution of the various oligonucleotides is roughly the same as that observed with the analytical 4.6 mm x 250 mm APS column with some exceptions. Specifically, the two pairs of oligomers ApApUpUpCpGp, ApUpUpUpApm^{2}Gp and CpUpCpApGp, m^{7}GpUpCpm^{5}CpUpGp which eluted as two sets of fused peaks from the analytical APS column (figure 10) eluted as two single peaks on the preparative column. The two peaks were collected (each containing two oligomers), desalted and subsequently chromatographed on analytical reverse-phase column (ODS). The analysis of the mixture containing CpUpCpApGp and m^{7}GpUpCpm^{5}CpUpGp on the analytical ODS column is shown in the first chromatogram of figure 13. This analysis again indicates the advantage of using both anion-exchange and reverse-phase

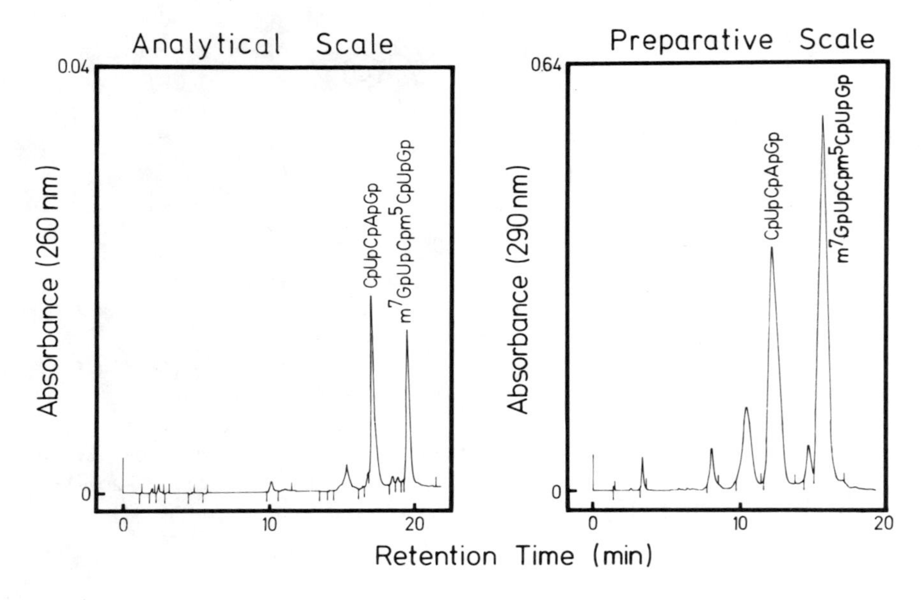

Figure 13 Chromatography of the oligonucleotide mixture containing CpUpCpApGp and m^7GpUpCpm^{5}CpUpGp on Zorbax®-ODS. Analytical scale: 0.3 A_{260} units on a 4.6 mm x 250 mm column. Solvent A: 0.05 M KH_2PO_4, pH 5.5. Solvent B: 70% CH_3OH in 0.05 M KH_2PO_4, pH 5.5. Flow: 2.0 mL/min; temperature: 35°C; detector: 260 nm, 0.04 aufs; gradient: 0-25% solvent B, 30 min. Preparative scale: 400 A_{260} units on a 21.2 mm x 250 mm column. Column conditions as for analytical scale separation with following exceptions, flow: 40 mL/min; detector: 290 nm, 0.64 aufs

chromatography for oligonucleotide isolation. These two oligomers which had coeluted on the anion-exchange column (APS) are now well resolved on the reverse-phase column (ODS). Additionally other minor contaminating constituents are visible. The separation was then scaled up from the analytical to the preparative ODS column using exactly the same gradient conditions. The flow rate through the column was increased by a factor representing the ratio of the amount of stationary phase in the preparative column to that of the analytical column (about 21). Upon injection of 400 A_{260} units of the oligonucleotide mixture the second chromatogram of figure 13 was produced. It is noteworthy that the preparative and analytical separations are comparable both in resolution and retention times. The ratio of the peak areas has changed which reflects the differing extinction coefficients at 290 nm as compared with 260 nm. After desalting and concentration 170 A_{260} units of $m^7GpUpCpm^5CpUpGp$ and 150 A_{260} units of CpUpCpApGp were obtained. A similar separation and isolation also occurred with the two oligonucleotides ApApUpUpCpGp and $ApUpUpUpApm^2Gp$.

While it is more convenient in scaling up separations to be able to increase the flow rate as high as 40 mL/min using a preparative hplc pump head, preparative hplc columns function adequately using analytical pumps with flow rates in the 8-10 mL range.

5 CONCLUSIONS

hplc is and continues to become a powerful technique for the isolation and purification of oligonucleotides. While we presently advocate the use of both anion-exchange and reverse-phase modes of chromatography in this respect, it could be that new developments in stationary phases which combine both selection principles, or modifications of the mobile phase, such with ion pair partitioning, will result in enhanced resolution of oligonucleotide mixtures.

6 INSTRUMENTATION

The chromatograms used in the applications section of this chapter were performed on: DuPont 8823 gradient system including heated column compartment, uv-spectrophotometric detector with analytical or preparative cell. Preparative hplc was done using the above system with a preparative pumphead.

® registered trade mark of DuPont

REFERENCES

[1] Howard, G.A. and Martin, A.J.P. (1950) Biochem. J. 56, 539

[2] Halasz, I. and Sebastian, I. (1969) Angew. Chem. 8, 453

[3] McLaughlin, L.W., Cramer, F., and Sprinzl, M. (1981) Anal. Biochem. 112, 60

[4] Majors, R.E. (1977) J. Chromatog. Sci. 15, 334

[5] Gait, M.J. (1980) in "Polymer-supported reactions in organic synthesis", ed. P. Hodpe and D.C. Sherrington, John Wiley, p. 435

[6] Tyson, R.W. and Wickstrom, E. (1980) J. Chromatog. 192, 485

[7] Bischoff, R. and McLaughlin, L.W., unpuplished results

[8] Hjerten, S. (1976) "Methods of Protein Separation", N. Catsimpoolas, ed., Vol. 11, 233

[9] Matteucci, M.D. and Caruthers, M.H. (1981) J. Amer. Chem. Soc. 103, 3185

[10] Gumport, R.I. and Uhlenbeck, O.C. (1981) in "Gene Amplification and Analysis. Analysis of Nucleic Acid Structure by Enzymatic Methods", I.G. Chirikjian and T.S. Papas, eds., Vol. II, Elsevier North Holland, New York, in press

[11] Uhlenbeck, O.C. and Gumport, R.I. (1981) "The Enzymes", 3rd ed., Vol. 14, in press

ANALYSIS OF SYNTHETIC OLIGODEOXYRIBONUCLEOTIDES

Hans-Joachim Fritz, Dirk Eick and Wolfgang Werr

Institut für Genetik
Universität zu Köln

SUMMARY

High pressure liquid chromatography has become a standard method in chemical DNA synthesis and represents the ideal complement to the solid phase synthesis of oligodeoxynucleotides. Since the product of polymer supported oligonucleotide synthesis typically is a fairly complex mixture of compounds, final success critically depends on the analytical part of the work. This article deals with the identification and purification of oligonucleotides by chromatography on C_{18} columns in the paired ion mode as an equivalent to anion exchange chromatography. It furthermore gives experimental details for determination of the base composition and identification of the 5'-terminus of a synthetic DNA fragment. Modifications of the Maxam Gilbert procedure for DNA sequence analysis by chemical degradation are described, which allow its convenient application to short (synthetic) oligodeoxyribonucleotides.

Alternative combinations of methods for purification and complete characterisation of oligodeoxyribonucleotides (all resting on either reversed phase hplc or polyacrylamide gel electrophoresis) are discussed.

1 INTRODUCTION

During the last few years, synthetic oligonucleotides have become widely used tools for molecular biologists and genetic engineers; the demand of these compounds is presently in a phase of rapid growth.

On the one hand, the newly emerging molecular cloning technology has made it a simple task to shuttle synthetic DNA between the test tube and living cells, whereby the DNA can be purified, amplified and studied with respect to its biological properties (for two early examples of cloning of synthetic DNA see [1,2]).

On the other hand, recent progress in chemical methodology has dramatically decreased the amount of time and effort required for oligonucleotide synthesis. Preparation of synthetic DNA no longer needs to be the limiting factor of a biological experiment, that can profit from such material.

This progress was brought about by the combined improvements in three different areas:

- Fast, yet selective phosphorylation procedures have been developed (phosphotriester [3] and phosphite [4,5] method).
- For solid phase synthesis of oligonucleotides, practical insoluble supports have been devised, that are compatible with the above mentioned phosphorylation methods [5-9].
- High pressure liquid chromatography (hplc) proved to be a very powerful method for both preparative purification and analysis of oligonucleotides [10,11].

While chemical DNA synthesis used to be the domain of a few specialised laboratories, solid phase technology with its ease of handling and its built-in option of process automation will clearly help to proliferate the method throughout the realm of molecular biology.

The convenience of polymer supported synthesis, however, has its price: the synthetic intermediates remain attached to the insoluble matrix during the entire process of chain assembly and are thus not amenable to purification. Only after the last chain elongation step is the product released from the polymer, at which time it has undergone numerous cycles of phosphorylation and terminal deprotection, each cycle possibly being combined with terminal blocking ("capping") and oxidation of trivalent phosphorus, depending on the synthetic strategy applied.

None of these reactions is truly perfect with respect to turnover and selectivity, and, in addition, it is known, that the final chemical manipulations for complete deblocking of the product are accompanied by side-reactions.

It is, therefore, not surprising, that, typically, the product of a polymer supported oligonucleotide synthesis is a fairly complex mixture of compounds.

This, in turn, shifts a good part of the burden from synthesis to the subsequent analytical work, in which three problems have to be solved:

- Within the mixture of compounds, the desired product has to be identified.
- This product has to be purified from the various contaminants.
- The purified product has to be analysed with respect to homogeneity and chemical identity.

Recently developed methods of liquid chromatography (see above) have helped to meet these requirements and have thus contributed to the success of solid phase oligonucleotide synthesis.

This article briefly deals with the first two points and in greater detail with the third one; emphasis is on hitherto unpublished work.

The methods described are selected in such a way as to make extensive use of only two simple separation procedures: polyacrylamide gel electrophoresis (page) and reversed phase hplc; the aim is to enable even laboratories with no special experience in synthetic DNA chemistry to purify and fully characterise "home-made" and commercial oligodeoxynucleotides.

2 REVERSED PHASE hplc OF UNDEGRADED OLIGODEOXYRIBONUCLEOTIDES

2.1 Introduction

High pressure liquid chromatography has become a standard method chemical DNA synthesis. It can be used for both analysis and preparative purification of oligonucleotides.

Technical aspects of the method as well as basic concepts of a structural rationale behind the chromatographic behaviour of oligonucleotides on both reversed and anion exchange phases are amply dealt with in previous publications [10,11] and in other articles within this volume (see contributions by M.J. Gait and by L. McLaughlin and J. Krusche).

Therefore, we will restrict ourselves in this section to a brief discussion of the flexible use of only one type of stationary phase: the microparticulate, bonded C_{18} column.

2.2 Chromatography on C_{18} columns in the ordinary mode

High pressure liquid chromatography of oligonucleotides on reversed phase is characterised by the following features:

- High resolution, sensitivity and speed.
- Clear-cut correlation between the lipophilic character of the sample and its chromatographic behaviour.
- Straightforward scale-up of analytical runs by factors of up to 10^6 using the same instrument.

These characteristics make the method very useful for the following applications in connection with solid phase synthesis:

- Quick survey on homogeneity of crude final product (partly and completely deprotected).
- Preparative purification with good resolution and recovery: Even compounds of very similar structures are efficiently separated, e.g. oligonucleotides with supposedly identical sequences but differing in a single inadvertant structural deviation such as a protecting group which has escaped removal or a modified base created by a side reaction.
- Test of homogeneity of final (purified) product.

Since it is not always easy to predict the lipophilic character and hence the chromatographic behaviour of a given oligonucleotide, the method is less successful in product identification.

Unequivocal identification of the desired product obviously is the most important analytical task before setting out for preparative purification and can be a major problem if one is dealing with complex mixtures.

In the case of polymer supported synthesis, the desired product can safely be expected to be in the group of compounds with the greatest chain length; a method that separates predominantly according to this parameter is clearly more useful in this situation than "normal" C_{18} chromatography.

Strong anion exchange hplc has successfully been used for product identification and (partial) purification from crude mixtures (see contribution to this volume by M.J. Gait).

We suggest that chromatography on C_{18} columns in the "paired-ion" mode might be used equivalently to anion exchange chromatography (see below).

2.3 Chromatography on C_{18} columns in the "paired-ion" mode

Deprotected oligonucleotides are polyanions. Their retention behaviour on C_{18} columns varies with the nature of cations present in the eluent: At identical concentrations of acetonitrile, a given oligonucleotide elutes more slowly with 0.1 M triethylammonium acetate as the eluent than with 0.1 M ammonium acetate [10].

This finding is in agreement with the concept of paired-ion chromatography [12]. The effect of increased retention in the presence of more lipophilic counterions has been explained in two formally different ways [13]:

- Negative charges of the sample molecules are neutralised in solution by cations present in the eluent. The sample molecules enter the lipophilic surface on the stationary phase as neutral entities, the lipophilicity of which is determined i.a. by the associated counterions.
- The lipophilic cation itself is retained by the C_{18} surface giving it anion exchange properties.

Both models explain equally well why with highly lipophilic cations, such as tetrabutylammonium, the separation of oligonucleotides on C_{18} columns is essentially by increasing chain length (see below).

There might be, in fact, no molecular justification to distinguish between the two models: both invoke a ternary association "complex" as being responsible for the affinity of the sample to the stationary phase:

sample (poly)anion (A) / lipophilic cation(s) (B) / C_{18} surface (C)

The two models seem to differ only in two alternative ways of formally and arbitrarily breaking this complex up in subgroups of (A+B) and (C) or (A) and (B+C).

In this and in similar cases (such as RPC-5 chromatography [14]) a useful operational way of discriminating between reversed phase and anion exchange chromatography, which does not require speculations on molecular mechanisms, could be by regarding the nature of the eluent: is the strength of the eluent increased by addition of organic solvent or by increasing the salt concentration?

With this criterion applied, C_{18} chromatography in the paired-ion mode, as used here, is a reversed phase technique, RPC-5 chromatography an anion exchange method.

Figure 1 shows an example of a chromatogram of an undecanucleotide on a C_{18} column in the paired-ion mode (also compare figure 4).

In similar experiments (data not shown), shorter oligonucleotides eluted faster from the column than the undecanucleotide, longer ones more slowly. A second undecamer, with markedly different elution behaviour on C_{18} in the ordinary mode, eluted with the same retention time in the paired-ion experiment.

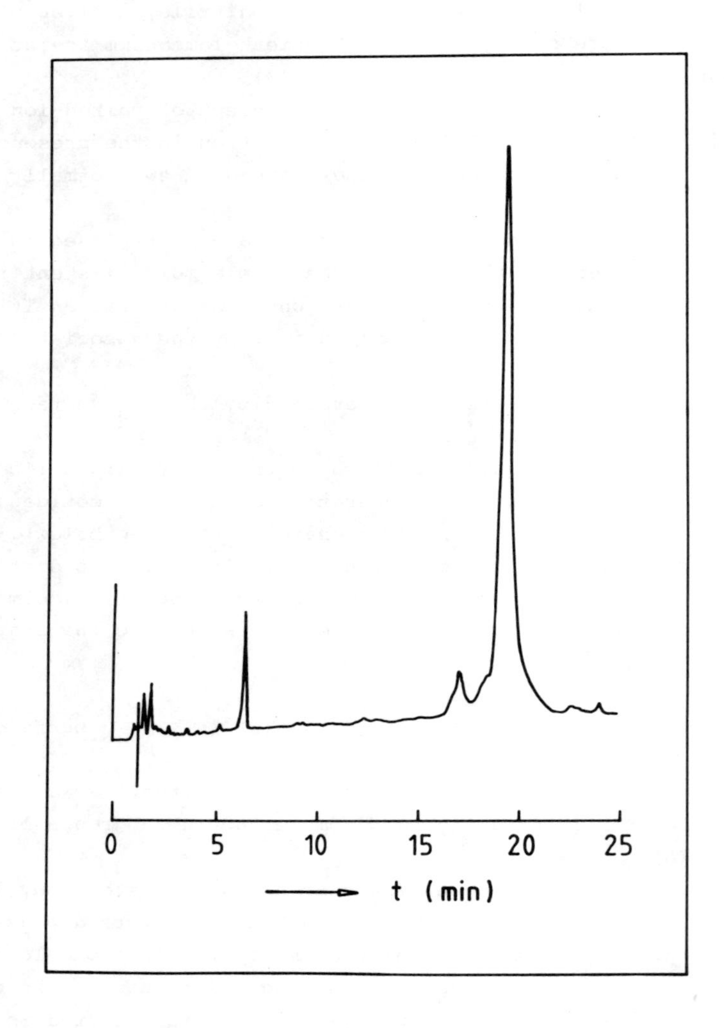

Figure 1 Reversed phase hplc tracing of the undecanucleotide dAATTCATGTGT. Chromatographic conditions: stationary phase: Zorbax-ODS (DuPont). Mobile phase: eluent A: water containing 5 mM tetrabutylammonium phosphate, pH 7.5; eluent B: 20% water, 80% acetonitrile containing 5 mM tetrabutylammonium phosphate, pH 7.5 (WATERS PIC-A); linear gradient from 55% B to 90% B in 30 min. Detection: uv-absorption (0.16 AUFS at 260 nm). Flow: 2 $mL \cdot min^{-1}$.
The material was synthesised by EMBO-course students under the supervision of Dr. B. Kaplan. The chromatogram was recorded and kindly provided by Dr. J. Krusche

All this strongly suggests that chain length is the predominant factor in these separations; the body of experience, however, is too thin at present for a sound evaluation of the method. Systematic studies are needed now for direct comparison of paired-ion chromatography and anion exchange chromatography of oligonucleotides.

3 BASE COMPOSITION ANALYSIS BY FORMIC ACID HYDROLYSIS FOLLOWED BY QUANTITATIVE REVERSED PHASE hplc

3.1 Introduction

Treatment of DNA with 90% formic acid cleaves the glycosidic bonds to liberate the free nucleobases; the remaining sugar-phosphate backbone is degraded [15,16]. It has been shown, that under conditions of complete hydrolysis less than 1% each of adenine, cytosine and 5-methylcytosine are converted into the respective deamination products [17].

The base composition of a DNA preparation can thus be determined by separation of the nucleobases and quantitative evaluation of the fractions obtained.

Classical methods for this type of analysis are paper chromatography and ion exchange chromatography on the one hand and uv-spectroscopy and isotope dilution techniques on the other [16-19].

In recent years, high pressure liquid chromatography of nucleic acid components on reversed phase [10] proved to be successful in a variety of different applications. Direct quantitative analysis, however, of base composition by reversed phase hplc of DNA hydrolysates is somewhat impeded by the fact, that recovery of nucleobases from ordinary C_{18} columns does not exactly reflect the input amounts, due to tight adsorption of parts of the sample to the stationary phase [20]. We found the loss of material to be in the order of up to 10 to 20% for the different nucleobases.

In this section we describe a fast and accurate method of determining the base composition of synthetic oligodeoxynucleotides which rests on formic acid hydrolysis plus reversed phase hplc and which overcomes the above mentioned adsorption problem by calibration using a standard DNA of known nucleotide sequence: plasmid pBR322 [21].

Advantages of this type of calibration are twofold:

No weighing out of reference material with its intrinsic inaccuracies is required.

There is no need to determine the molar extinction coefficients of the nucleobases under the conditions of the experiment.

3.2 Procedures

3.2.1 Hydrolysis of reference DNA

50 to 100 µg of reference DNA (pBR322 or another DNA of known nucleotide sequence) are precipitated with two volumes of ethanol in the presence of 0.1 M NaCl. The pellet is dried with N_2. Traces of RNA are removed by hydrolysis in 100 µL 1 N NaOH at 50°C for 1 h. Hydrolysis is stopped by adding 100 µL 1 M Tris-HCl, pH 7.5 and 100 µL 1 N HCl. Subsequently, the DNA is precipitated by addition of two volumes of ethanol at room temperature. A second precipitation is made from 100 µL distilled water, 10 µL 3 M NaCl and 220 µL ethanol. The DNA is dried again with N_2 and the pellet is redissolved in 200 µL 90% formic acid.

Hydrolysis of the DNA is performed at 170°C in a 4 mL pyrex tube fitted with a screw-cap sealed by a rubber-teflon sandwich gasket. After 30 min of reaction, the heating is switched off and the sample remains in the oven until room temperature is reached. Subsequently, the sample is frozen at -80°C, then lyophilised.

After the reaction, the tubes should be handled with care (gloves, goggles) and not be opened at room temperature (possible carbon monoxide pressure).

3.2.2 Hydrolysis of synthetic oligodeoxynucleotides

The procedure for hydrolysis of synthetic oligonucleotides is the same as for marker DNA, with the exception, that no precipitation and sodium hydroxide pre-treatment is required. Rather, an aqueous solution of ca 50 µg of oligonucleotide is directly mixed with 9 volumes of pure formic acid and heated for hydrolysis.

3.2.3 Chromatography of hydrolysates

For hplc measurements, 50 to 100 µg of hydrolysed DNA are redissolved in 50 to 100 µL of the following buffer: 0.1 M NH_4OAc:acetic acid (75:25, v/v). Aliquots of 5 to 10 µL are injected onto the column. Chromatographic conditions are as given in figure 2.

It is important to mention, that all measurements must be done under identical conditions to guarantee comparable quantitative results. This means for the optical detection unit, that measurements of the sample and of the reference hydrolysate should be done at the same sensitivity setting; for the integrator, that registration of peaks should be done with identical slope and width sensitivities.

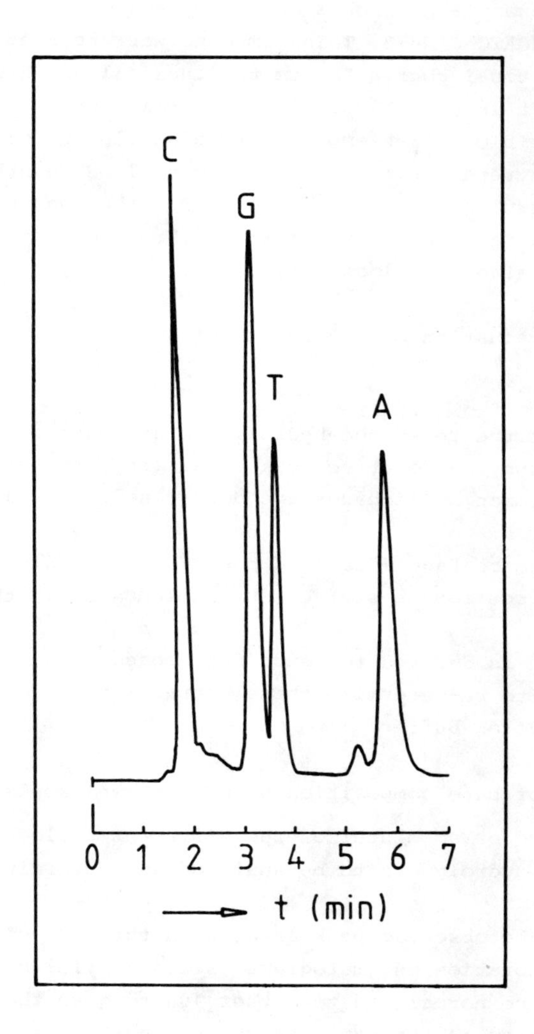

Figure 2 Reversed phase hplc tracing of formic acid hydrolysate of the undecanucleotide dGATCACACATG.
Chromatographic conditions: stationary phase: 0.39 cm x 30 cm µ-Bondapak-C_{18} (WATERS). Mobile phase: 0.1 M ammonium acetate, pH 4.25, containing 1.5% acetonitrile. Flow: 2 mL·min^{-1}. Detection: uv-absorption (0.01 AUFS at 280 nm). Elution times: 1.73 min (C), 3.23 min (G), 3.67 min (T), 5.23 min (impurity) and 5.80 min (A)

Figure 2 shows an example of such a chromatogram: we have analysed the undecanucleotide dGATCACACATG. This compound was prepared by students of the EMBO practical course "Automated Chemical and Enzymic Gene Synthesis" in Darmstadt in April 1982. A phosphotriester strategy using preformed dinucleotide blocks was employed to assemble the oligonucleotide chain on a polystyrene resin in a BACHEM manual DNA synthesizer. The synthesis was supervised by Dr. B. Kaplan. For the analysis we used a "semi-purified" product, i.e. no preparative chromatography had been applied to the sample after final detritylation.

3.2.4 Quantitative evaluation of the chromatogram

3.2.4.1 Calibration

Several aliquots of the reference hydrolysate are chromatographed separately. For each run the relative peak areas (in percent) corresponding to the four common nucleobases are determined. The values are averaged.

The molar residue percentage ("base composition") is calculated from the known numbers of occurrences of the four nucleobases in the reference DNA.

A correction factor is derived for each nucleobase as shown in table 1. We recommend to re-determine these correction factors for every new batch of elution buffer.

3.2.4.2 Calculation of base composition of an unknown sample

Peak areas (relative or absolute) of the different nucleobases in the chromatogram of a hydrolysate to be analysed are determined as described above.

Multiplication of the observed peak areas with the correction factors derived from calibration chromatograms (3.2.4.1) yields corrected peak areas, which can be normalised to either 100 to give the base composition in percent or to the expected chain length of the oligonucleotide to give the number of residue equivalents in the compound.

The procedure is summarised in table 2.

Table 1 Calculation of correction factors for peak areas from calibration chromatograms with pBR322 hydrolysate (measured at 280 nm)

Base	Observed relative peak areas (average of 8 measurements)	Number of residues in pBR322 +)	Molar residue percentage	Correction factor ++)
C	38.89	2331	26.91	0.692
G	35.26	2343	27.05	0.767
T	14.16	2018	23.30	1.645
A	11.68	1970	22.74	1.947
Σ:	99.99	8662	100.00	

+) 12 C-residues and 48 A-residues are subtracted from the gross base composition, to account for known E. coli K12 methylation sites.

++) Obtained by dividing the molar residue percentage by the observed relative peak area.

Table 2 Quantitative evaluation of the chromatogram shown in figure 2

Base	Elution time [min]	Peak area	Correction factor +)	Corrected peak area	Corrected peak area normalised to 100% ++) [%]	Corrected peak area normalised to 11 Res. ++)
C	1.73	23984	0.692	16600	28.78 (27.27)	3.17 (3)
G	3.23	12626	0.767	9684	16.79 (18.18)	1.85 (2)
T	3.67	6852	1.645	11271	19.54 (18.18)	2.15 (2)
A	5.80	10335	1.947	20122	34.89 (36.36)	3.84 (4)
			Σ:	57677	100.00 (99.99)	11.01 (11)

+) see table 1

++) expected values are given in parentheses

3.3 Discussion

A) The analysis we chose as an example, though performed on a semi-purified product of a solid-phase synthesis, predicts the correct base composition, assuming the compound is an undecanucleotide. This means, that the method can be regarded as a quick and convenient way to check the identity of an isolated synthetic product.

B) The method will be less helpful in assessing product homogeneity, since - depending on nucleotide sequence and synthetic strategy - contaminating oligonucleotides (individually or as a population) may have similar or identical base composition as the desired product. In particular and for obvious reasons, this type of analysis will fail to detect chain degradation as it may occur during the final deprotection steps.

C) The method is in principle suitable for analysis of base modifications, that have occurred during synthesis - provided the modified bases are stable under the conditions of the formic acid treatment (or at least lead to identifiable products). We regard the small peak at 5.23 min in figure 2 as a likely candidate for being caused by such a modified base: this material is not present in the calibration hydrolysates and is, therefore, not an artefact of the formic acid treatment. We have not yet followed up this aspect of the method any further, but we expect, that with more data being accumulated, it will contribute to our understanding of undesired side-reactions in chemical DNA synthesis.

D) We find peak areas of cytosine somewhat unreliable. This is due to the fact, that under standard chromatographic conditions, this base has very little affinity to the stationary phase and co-elutes with small amounts of other uv-absorbing material (condensation products of 2'-deoxyribose?). We are presently trying to improve this situation by paired-ion chromatography. In a preliminary experiment we found WATERS PIC B-5 reagent in water, used at half the concentration recommended by the supplier, to be helpful: under otherwise identical conditions as used for the chromatogram shown in figure 2, retention times are as follows: 3.70 min (C), 4.34 min (T), 5.26 min (G) and 7.03 min (A).

4 COMBINED ANALYSIS OF BASE COMPOSITION AND 5'-TERMINAL RESIDUE BY TOTAL SNAKE VENOM PHOSPHODIESTERASE DIGESTION FOLLOWED BY REVERSED PHASE hplc IN PAIRED-ION MODE

4.1 Introduction

Snake venom phosphodiesterase degrades linear polynucleotides with 3',5'-phosphodiester linkages in an exonucleolytic fashion from the 3'-end, liberating nucleoside-5'-monophosphates [22]. Non-phosphorylated 5'-terminal residues (as normally present in synthetic oligonucleotides) are left behind as free nucleosides. The reaction product of a snake venom phosphodiesterase digest of any terminally non-phosphorylated oligonucleotide made up of "normal" DNA constituents and containing 3',5'-phosphodiester linkages exclusively will, therefore, consist of a mixture of one nucleoside and one to four nucleoside-5'-monophosphates. Such a mixture can easily be analysed by reversed phase hplc in the paired-ion mode as demonstrated below.

4.2 Procedure

60 µg of fully deprotected oligodeoxynucleotide are digested with 0.5 units of snake venom phosphodiesterase (Worthington) in 100 µL 0.1 M Tris-HCl, pH 9, 0.1 M NaCl, 15 mM $MgCl_2$ at 25°C.

Aliquots of 10 µL each are removed from the reaction mixture after 30 min, then every full hour. The aliquots are sealed immediately in glass capillary tubes (for instance 50 µL disposable micropipettes) and kept at 95°C for 15 min.

Each aliquot is then analysed by hplc. Chromatographic conditions are as follows: stationary phase: 0.39 cm x 30 cm µ Bondapak-C_{18} (WATERS) or equivalent reversed phase column; mobile phase: 5 mM tetrabutylammonium phosphate, pH 7.5 (for instance WATERS Pic A reagent) containing 15% acetonitrile; flow: 2 $mL \cdot min^{-1}$.

When no more change in the elution pattern is observed, the reaction is stopped by heating (see above). The mixture is then examined by hplc as described above, but with only 4% acetonitrile present in the eluent. A homogeneous oligonucleotide will exhibit one peak in the "nucleoside region" of the chromatogram (1 to 5 min) and up to four peaks in the "nucleoside-monophosphate region" (5 to 20 min).

Integrated peak areas (detection at 254 nm) are divided by the following correction factors ($\varepsilon_{254} \times 10^{-4}$, interpolated from a table of molar extinction coefficients in use in the laboratory of Dr. H.G. Khorana): 1.40 (A), 0.64 (C), 1.36 (G) and 0.64 (T). Ratios of corrected peak areas equal molar ratios of different compounds present in the digestion mixture.

Complete digestion of the oligonucleotide is confirmed by chromatography of the digestion mixture using a gradient of acetonitrile from 0 to 80% in 90 min (other conditions as described above).

Figure 3 shows the result of such an analysis, obtained from enzymatic hydrolysis of the octanucleotide dGAAGCTTC, which was synthesised by phosphotriester methods in solution [23].

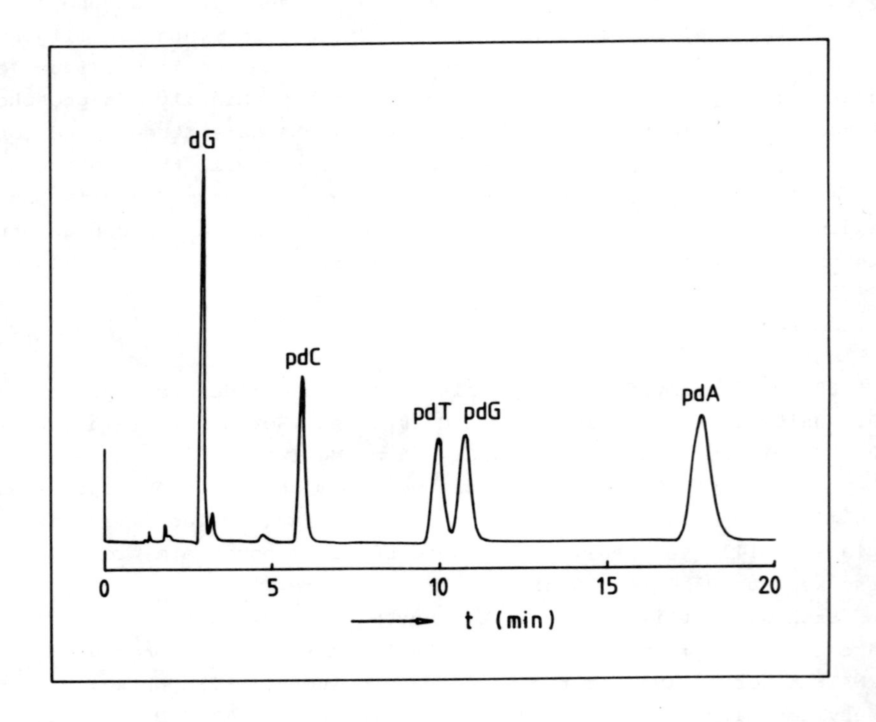

Figure 3 Reversed phase hplc tracing of products obtained from complete enzymatic digestion of octanucleotide dGAAGCTTC with snake venom phosphodiesterase. Mobile phase: water containing 5 mM tetrabutylammonium phosphate, pH 7.5 (WATERS PIC A) and 4% acetonitrile. Detection: uv-absorption (0.02 AUFS at 254 nm). For other conditions refer to text. Peak integration plus correction for different extinction coefficients (see text) yielded the following monomer composition (normalised to a sum of 8 residues): dG: 1.1 (1), pdG: 1.08 (1), pdC: 1.98 (2), pdT: 1.94 (2) and pdA: 1.90 (2); expected values are given in parentheses

4.3 Discussion

The following types of information can be derived from the analysis of a complete snake venom phosphodiesterase digest of a synthetic oligonucleotide:

A) The 5'-terminal residue is identified. If the sample was purified from a complex mixture (as it is often the case in polymer supported synthesis), this test can in many instances be helpful to clarify the question if the isolated compound is a product of the final chain elongation reaction or if it represents a truncated sequence. Use of the test in this way is limited, of course, to synthetic schemes with 3'- to 5'-chain elongation, as in the phosphotriester and phosphite methods.

B) The homogeneity of the 5'-terminus can be estimated. A note of caution, however, has to be added at this point: contaminating and/or intrinsic phosphatase activity of commercial preparations of snake venom phosphodiesterase can interfere with the analysis, a problem, that can - at least partly - be overcome by a special pre-treatment of the enzyme [24].
In any case, the reaction should be optimised to the shortest possible time necessary for complete digestion, in order not to expose the liberated nucleotides to this phosphatase activity unduly long.

C) Base composition of the oligonucleotide can be calculated. For the information content of this analysis the same arguments apply as for the formic acid procedure (paragraph 3).

D) (Mean) chain length of the oligonucleotide (mixture) can be estimated by dividing the entire corrected peak area by the corrected peak area of the material eluting in the nucleoside region of the chromatogram. With increasing chain length, however, this analysis becomes more and more inaccurate, since it involves dividing by a relatively small number.

E) Exclusive presence of "natural" 3',5'-phosphodiester linkages in the product is confirmed, if the sample is completely degraded to nucleoside(s) and nucleoside-5'-monophosphates.
This is perhaps the most important single piece of information, that can be obtained by snake venom phosphodiesterase digestion.

5 Sequence analysis of synthetic oligodeoxynucleotides by the chemical degradation method

5.1 Introduction

The two preceeding paragraphs describe fast and convenient tests of product identity of somewhat limited information content. While these tests might be regarded sufficient in many instances, a truly complete analysis of an oligonucleotide has to incorporate sequence determination.

In principal, there are two straightforward methods available for sequence analysis of oligodeoxynucleotides:

- The "wandering spot" method originally developed by F. Sanger and colleagues (see ref. [25] and contribution to this volume by R. Frank and H. Blöcker).
- The chemical degradation method of A. Maxam and W. Gilbert [26].

For synthetic oligonucleotides, the former method has been used almost exclusively, since, as it has been pointed out [25], technical difficulties are associated with adaptation of the Maxam-Gilbert protocols to complete sequence analysis of short oligonucleotide chains.

In this section, we describe ways to overcome these difficulties and demonstrate successful application of the method to oligodeoxynucleotides of chain lengths down to eight.

Sequence analysis of synthetic oligonucleotides by the Maxam-Gilbert method has briefly been mentioned in literature [8]; however, no experimental details were given.

5.2 Procedures and discussion

In the following, the course of a sequencing experiment is described. Specific problems usually encountered with short oligonucleotides are pointed out as well as modifications of the standard procedures [26]. Where no special mention is made, the original protocols [26] are followed strictly.

5.2.1 Terminal labelling of oligodeoxynucleotides using [γ-^{32}P]ATP and polynucleotide kinase

A solution of ca 0.5 nmol of completely deprotected oligodeoxynucleotide (free of ammonium ions) is evaporated to dryness. The remainder is taken up in 10 µL of aqueous [γ-^{32}P]ATP (AMERSHAM or NEN; 5000-6000 Ci$\cdot$mmol^{-1}; 10 mCi$\cdot$mL^{-1}), 1.5 µL 10X kinase buffer (0.4 M

Tris-HCl, pH 9, 0.1 M $MgCl_2$, 50% glycerol). 1.5 µL 0.1 M dithiothreitol are added and the mixture is incubated with 2 µL (10 units) of polynucleotide kinase (Boehringer) at 37°C for 30 min.

The reaction is terminated by addition of 2 µL 0.1 M EDTA.

Co-precipitation of oligonucleotide with carrier DNA is carried out by addition of 5 µL sonified calf thymus DNA (1 mg•mL^{-1}, in 10 mM Tris-HCl, pH 8, 0.25 mM EDTA), 25 µL 3 M ammonium acetate, 203 µL water and 750 µL ethanol followed by standing at -20°C for 1 h and centrifugation (Sorvall SS 34 rotor, 15,000 rpm, 20 min, -20°C). The supernatant is removed and the pellet is dried, then taken up in 10 µL water for preparative polyacrylamide gel electrophoresis (see below).
Problem: Oligonucleotides of chain lengths below ca 15 show significant solubility in 75% ethanol. Loss of material in ethanol precipitations can be kept minimal by generous use of carrier nucleic acid and by keeping the temperature at -20°C until the supernatant is removed. In addition, the number of ethanol precipitations should be kept minimal.

5.2.2 Preparative polyacrylamide gel electrophoresis of terminally labelled oligonucleotides

Polyacrylamide gel electrophoresis of terminally labelled oligonucleotides is carried out as described [26].

It serves the following purposes:

1. Removal of residual [γ-^{32}P]ATP.
2. Purification of major product from potential oligonucleotidic contaminants.

Figure 4 gives an example of patterns obtained from polyacrylamide gel electrophoresis.

Oligonucleotides are recovered from gels the following way: Small rectangles of film depicting the major bands are cut out from one of the two X-ray films obtained from the procedure described in the legend to figure 4. Making use of the radioactive ink markers for alignment, this film is then used as a template for the excision of gel slices containing the major bands. Individual gel slices are transferred to Eppendorf tubes, then crushed and finely ground after addition of 0.5 to 0.8 mL water or TE buffer (10 mM Tris-HCl, pH 8, 0.25 mM EDTA). The mixtures are kept at 65°C for 10 to 20 h with occasional shaking. The supernatants are removed after brief centrifugation; the pellets are washed once with 250 µL TE buffer each. The isolated liquid phases are made up to 1 mL by addition of carrier DNA (8 µL, 1 mg·mL^{-1}; compare preceeding paragraph) and TE buffer, mixed and centrifuged. Four aliquots of 225 µL are carefully removed from

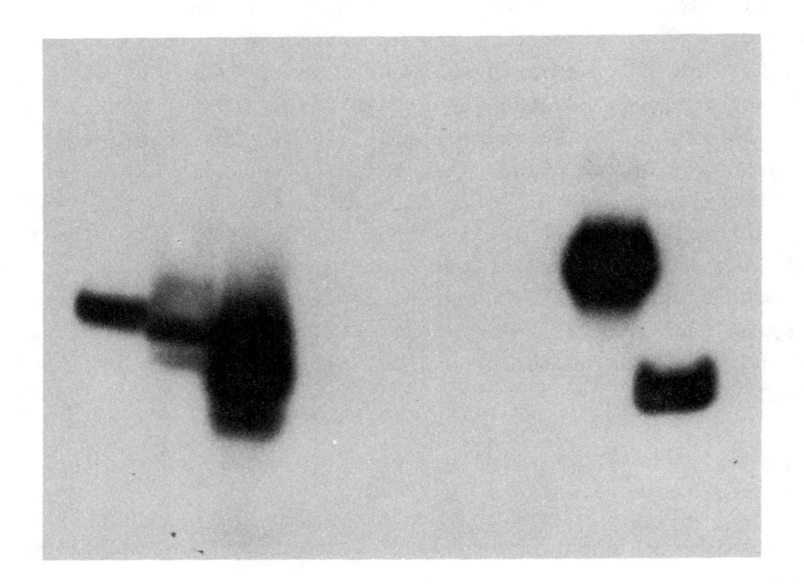

Figure 4 Polyacrylamide gel electrophoresis patterns obtained from 5'-terminally [^{32}P]-labelled oligonucleotides (1/2-aliquots of polynucleotide kinase reaction products, see text). Lane A: d(A-A-T-T-C-A-T-G-T-G-T) (compare figure 1), lane B: d(G-A-T-C-A-C-A-C-A-T-G) (compare figure 2), lane C: d(C-C-G-A-T-A-T-C-G-G), lane D: d(T-G-G-T-C-A-T-A-G-C-T-G-T-T-T-C-C-T-G), lane E: d(G-A-A-G-C-T-T-C) ([23], compare figure 3). The oligonucleotides leading to patterns C and D were synthesised by EMBO course students under supervision of Dr. M.J. Gait. Lanes C and D are heavily overexposed. Electrophoresis conditions: Gel: 0.15 x 20 x 20 cm; 20% (acrylamide:"bis" = 29:1), buffer: 0.178 M Tris base, o.178 M boric acid, 5 mM EDTA. Voltage: 200 V. Electrophoresis was cóntinued until the bromophenol blue marker (nót shown) had migrated ca 7 cm from the top.
Bands were made visible by autoradiography on two sheets of X-ray film simultaneously after wrapping the gel in "Saran" foil and pasting labels to the four corners of the gel with markers drawn on them with radioactive ink

each sample with the pipet tip just dipping into the solution. The remaining 100 µL potentially contain small pieces of polyacrylamide gel and are discarded (caution: bits of gel will severely interfere with the sequencing procedure!). 25 µL 3 M ammonium acetate, then 750 µL ethanol are added to each aliquot. The precipitation is done as described above.

5.2.3 Chemical degradation reactions

Chemical degradation reactions are carried out as described [26] (G: dimethylsulfate, A>C: 0.1 N NaOH, C+T: hydrazine, C: hydrazine, NaCl). Reaction times are prolonged by a factor of 2 (G; C+T) or 4 (A>C, C) respectively as compared to the standard procedures [26]. Problem: In the following sequence gel (see below), any residual traces of salt in the samples will lead to massive distortion of bands in the region of the first few nucleotides. Extreme caution must, therefore, be taken in complete removal of supernatants from ethanol precipitations. In addition, we found it useful to wash the pellets from the final precipitations with 200 µL ethanol (-20°C) with mixing and subsequent centrifugation (-20°C).

5.2.4 Polyacrylamide gel electrophoresis of chemical degradation products ("sequence gels")

Polyacrylamide gel electrophoresis of chemical degradation products to produce "sequence ladders" is carried out as described [26] and further illustrated in figures 5 and 6. With each set of samples, a marker of [^{32}P] phosphoric acid is co-electrophoresed in a fifth lane for identification of the band corresponding to the 5'-terminal residue. When 20% gels are used, electrophoresis is stopped, when the bromophenol blue marker has migrated ca 15% of the total length of the gel. Gel thickness of ca 1.5 mm is to be preferred over the more commonly used 0.3 mm since band distortion due to residual salt seems to be less problematic with a greater cross-section of the gel (presumably due to better dilution of the salt within the gel).

Figures 5 and 6 show sequence patterns obtained from synthetic oligonucleotides.

Problems: 1. We found that sometimes a 5'-terminal G residue gives rise to a second band (see sequences A and C in figure 5). We have not yet identified the nature of this "artefactual" product and have not yet learned to reproducibly suppress its formation. Incorrect interpretation of the gel pattern, however, can easily be avoided: The distance between bands corresponding to residues 1 and 2 (starting at

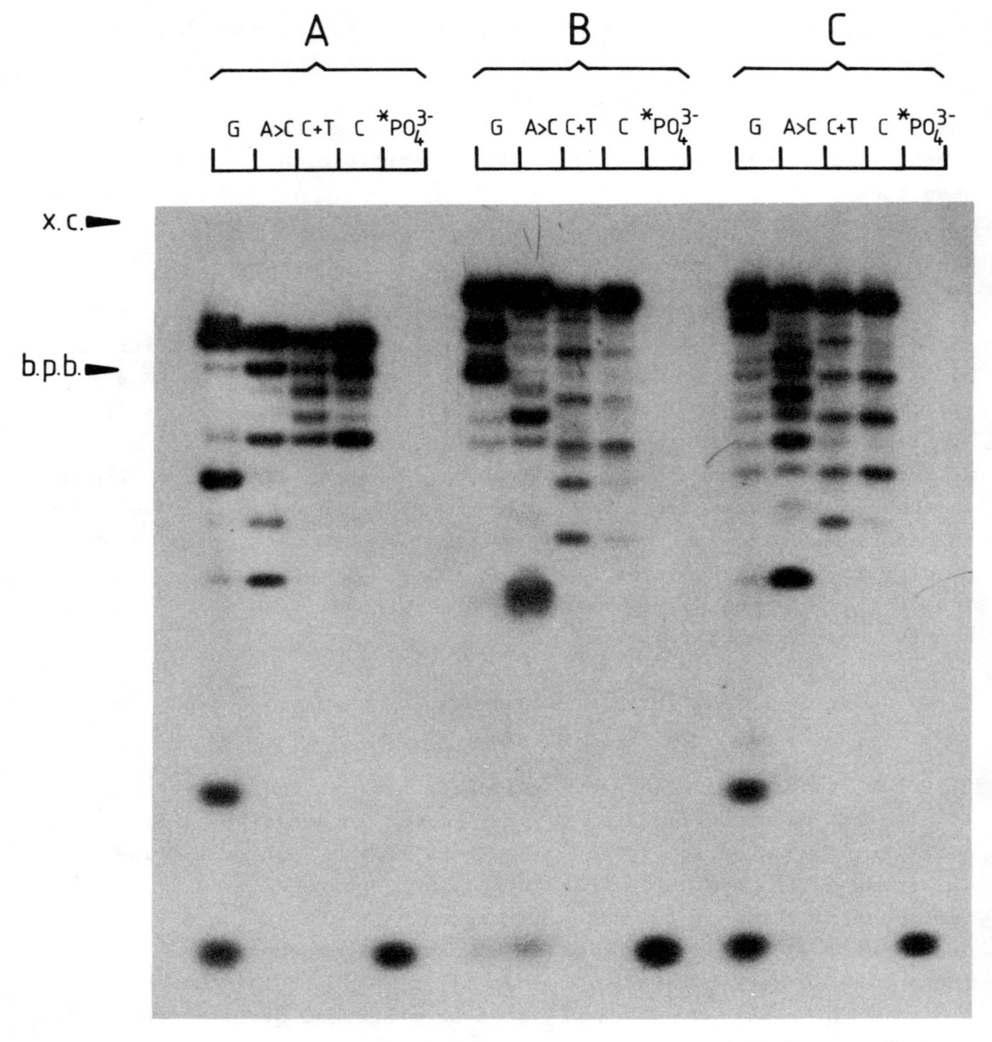

Figure 5 Sequence patterns obtained from synthetic oligodeoxynucleotides (refer to text for procedure). Pattern A: d(G-A-A-G-C-T-T-C) (compare figures 3 and 4), pattern B: d(A-A-T-T-C-A-T-G-T-G-T) (compare figures 1 and 4), pattern C: d(G-A-T-C-A-C-A-C-A-T-G) (compare figures 2 and 4). Electrophoresis conditions: Gel: 0.5 x 30 x 40 cm; 20% (acrylamide:"bis" = 19:1); buffer: 0.178 M Tris base, 0.178 M boric acid, 5 mM EDTA, 7 M urea. Voltage: 1100 V. Before sample loading, the gel was "pre-electrophoresed" until it had reached a temperature of ca 50°C. Autoradiography was carried out at -80°C to keep band broadening by diffusion minimal

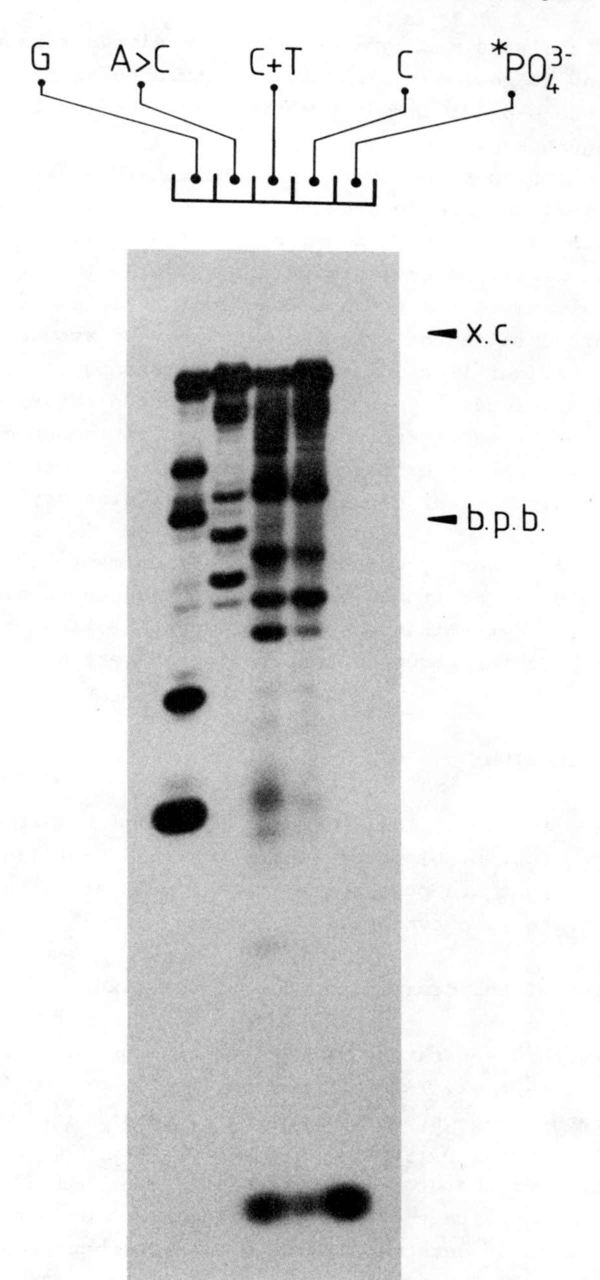

Figure 6 Sequence pattern obtained from synthetic d(T-G-G-T-C-A-T-A-G-C-T-G-T-T-T-C-C-T-G) (compare figure 4). Conditions were as described in the legend to figure 5

the 5'-terminus) is always much greater than the distance between bands corresponding to residues 2 and 3. With this being kept in mind, the upper of the two G-bands immediately reveals itself not to be part of the regular sequence pattern.

2. Bands corresponding to A residues are increasingly under-represented the closer these residues are located to the 5'-terminus. (See for instance 5'-terminal A-residue in sequence B, figure 5).

This effect can partly be compensated for by using more than one fourth of the total sample for the A>C reaction. This in turn, however, leads to over-representation of C in the A>C lane. We assume that the reason for this behaviour lies in the fact, that strong alkali is used in the A>C reaction, leading to some extent of chain scission. Short fragments would then be under-represented because of incomplete ethanol precipitation (in the other reactions, ethanol precipitations are done with intact chains; after cleavage with piperidine drying is done by evaporation).

We have not tried to verify our assumption experimentally, but for future experiments we think it may be more advantageous to use the G+A reaction [26] for determination of A-residues in oligonucleotides or whenever reading of the sequence pattern to the very 5'-terminus is desired.

6 SUMMARY AND CONCLUSION

How suitable are the analytical methods, described in paragraphs 2-5, to deal with the various problems of separation and identification (see Introduction), that are posed by a typical product mixture of solid phase oligonucleotide synthesis?

6.1 Identification of the desired product in the mixture

An analytical method separating (predominantly) by chain length is most suitable for this task. Two such methods are described in this article: A) Reversed phase chromatography on C_{18} stationary phase in the paired-ion mode and B) polyacrylamide gel electrophoresis.

The latter method can be used with radioactively labelled compounds and autoradiography for detection (as described here) or with "cold" material together with uv-detection (with optional enhancement of sensitivity by "uv-shadowing" - see contribution made to this volume by M.J. Gait et al.).

An alternative method is hplc on strong anion exchange stationary phase (see other contributions to this volume).

6.2 Preparative purification

All the methods described above lend themselves to scaling up to at least O.D. quantities. Among them, reversed phase chromatography is the one with the highest capacity and most straightforward sample recovery.

The material should best be purified by two subsequent procedures exploiting different separation principles.

Reversed phase chromatography in the "paired-ion" and the "normal" mode can be considered as two such independent methods.

6.3 Analysis of purified product

6.3.1 Product homogeneity

Product homogeneity is determined by analysing undegraded material. The same methods can be used as mentioned above and again, at least two independent tests should be carried out. Paragraph 2 of this article describes how this can be done using a single hplc column.

A powerful alternative method, yet technically more demanding and limited to detection of contaminations, that accept a radioactive label, is two-dimensional chromatography of undegraded material applying the conditions of the "wandering spot" sequencing method (see [25] and the article in this volume by R. Frank and H. Blöcker).

6.3.2 Product identity

Determination of product identity involves chemical or enzymatic degradation. Paragraphs 3-5 of this article describe three procedures that vary in speed and in content of information (qualitatively and quantitatively).

The "wandering spot" method (see above) is a well-established alternative to the Maxam-Gilbert procedure as described here for use with short oligonucleotides. The latter method, however, is technically less demanding and is also presently already in use in many biological laboratories, to which chemical oligonucleotide synthesis might be introduced as an additional technique.

Taken together, we feel confident, that the new and modified techniques described in this article, together with the literature cited, constitute a self-contained set of methods, which, by resting on only reversed phase hplc and polyacrylamide gel electrophoresis, should enable any typically equipped molecular biology laboratory to completely purify and analyse synthetic oligonucleotides, even if preparative methods were used, that yield complex product mixtures.

ACKNOWLEDGEMENTS

We thank Drs. M.J. Gait and B. Kaplan for gifts of material synthesized under their supervision by students of the EMBO Practical Course "Automated Chemical and Enzymic Gene Synthesis" (Darmstadt 1982). Some of the experiments described in this article were carried out under participation of students of the same course.

Work in the authors' laboratories was supported by grants made available to H.-J. F. and to W. Doerfler by Deutsche Forschungsgemeinschaft through SFB 74.

REFERENCES

[1] a) Khorana, H.G. (1979) Science 203, 614
b) Ryan, M.J., Belagaje, R., Brown, E.L., Fritz, H.-J., and Khorana, H.G. (1979) J. Biol. Chem. 254, 10803
[2] Fritz, H.-J. (1978) Nucl. Acids. Res., Special Publication No. 4, s243
[3] Reese, C.B. (1978) Tetrahedron 34, 3143
[4] Letsinger, R.L. and Lunsford, W.B. (1976) J. Amer. Chem. Soc. 98, 3655
[5] a) Matteucci, M.D. and Caruthers, M.H. (1981) J. Amer. Chem. Soc. 103, 3185
b) Beaucage, S.L. and Caruthers, M.H. (1981) Tetrahedron Lett. 22, 1859
[6] Gait, M.J., Singh, N., Sheppard, R.C., Edge, M.D., Greene, A.R., Heathcliff, G.R., Atkinson, T.C., Newton, C.R., and Markham, A.F. (1980) Nucl. Acids Res. 8, 1081
[7] Miyoshi, K. and Itakura, K. (1979) Tetrahedron Lett. 38, 3635
[8] Potapov, V., Veiko, V., Koroleva, O., and Shabarova, Z. (1979) Nucl. Acids Res. 6, 2041
[9] Crea, R. and Horn, T. (1980) Nucl. Acids Res. 8, 2331
[10] Fritz, H.-J., Belagaje, R., Brown, E.L., Fritz, R.H., Jones, R.A., Lees, R.G., and Khorana, H.G. (1978) Biochemistry 17, 1257
[11] Jones, R.A., Fritz, H.-J., and Khorana, H.G. (1978) Biochemistry 17, 1268
[12] Wittmer, D.P., Nuessle, N.O., and Haney, W.G. (1975) Anal. Chem. 47, 1422
[13] Waters Associates (1980) Product Information Pamphlet No. D61
[14] Hardies, S.C. and Wells, R.D. (1976) Proc. Natl. Acad. Sci. USA 73, 3117
[15] Wyatt, G.R. and Cohen, S.S. (1953) Biochem. J. 55, 774

[16] Loring, H.S. (1955) in The Nucleic Acids, Chargaff, E. and Davidson, J.N. Eds., Vol I, p. 191, Academic Press, New York

[17] Günthert, U., Schweiger, M., Stupp, M., and Doerfler, W. (1976) Proc. Natl. Acad. Sci. USA 73, 3923

[18] Cohn, W.E. (1955) in The Nucleic Acids, Chargaff, E. and Davidson, J.N. Eds., Vol I, p. 211, Academic Press, New York

[19] Wyatt, G.R. (1955) in The Nucleic Acids, Chargaff, E. and Davidson, J.N. Eds., Vol I, p. 243, Academic Press, New York

[20] Rustum, J.M. (1978) Analytical Biochemistry 90. 289

[21] Sutcliff, J.G. (1978) Cold Spring Harbor Symposia on Quantitative Biology 43, 77

[22] Khorana, H.G. (1961) in The Enzymes, Vol V, 2nd edition, Boyer, P.D., Lardy, H., and Myrbäck, K. Eds., p. 79, Academic Press, New York

[23] Werr, W. (1981) Part of "Diplom" Thesis, Universität zu Köln

[24] Sulkowski, E. and Laskowski, M. (1971) Biochem. Biophys. Acta 240, 443

[25] Tu, C.-P.D. and Wu, R. (1980) in Methods in Enzymology, Vol 65, Grossman, L. and Moldave, K. Eds., p. 620, Academic Press, New York

[26] Maxam, A.M. and Gilbert, W. in Methods in Enzymology, Vol 65, Grossman, L. and Moldave, K. Eds., p. 499, Academic Press, New York

THE "WANDERING SPOT" SEQUENCE ANALYSIS OF OLIGODEOXYRIBONUCLEOTIDES

Ronald Frank and Helmut Blöcker

GFB, Gesellschaft für Biotechnologische Forschung mbH
Braunschweig

SUMMARY

The experimental details of the two dimensional fingerprinting of oligodeoxyribonucleotides are described in detail. This method, also known as the "wandering spot analysis", allows a reliable sequence determination of oligonucleotides up to 40 nucleotides in length. Applied to synthetic oligonucleotides, this procedure is capable to prove whether the chemical synthesis has led to the wanted product free of the protective groups.

1 INTRODUCTION

1.1 Principle of the method

The "wandering spot" sequence analysis of an oligonucleotide is based on the characteristic mobility shifts of its sequential partial degradation products on a two dimensional chromatogram (2D-fingerprint) obtained by high voltage electrophoresis (1. dimension) and homochromatography (2. dimension) [1-3].

The oligonucleotide to be sequenced is labelled at its 5'-end using T4 polynucleotide kinase and [γ-^{32}P]ATP. It is then hydrolysed with snake venom phosphodiesterase (SVPD, EC 3.1.4.1). This enzyme cleaves off 5'-mononucleotides from the 3'-end of the oligonucleotide chain. During the course of hydrolysis aliquots of the reaction mixture are withdrawn and saved (see figure 1). From these aliquots a "cocktail" is prepared that contains all partial degradation products in equal amounts.

The "cocktail" is than subjected to high voltage electrophoresis on a cellulose acetate strip at pH 3.5. At this pH-value the difference in protonation of the 4 heterocyclic bases is maximal (see figure 2, table 1). As protonation partially neutralises the negative charges of the phosphate backbone the net charge of the oligonucleotide and hence its mobility in the electric field depends on its base content. This implies, that the difference in electrophoretic mobilities of two consecutive degradation products (n and n-1) is specified by the nature of the nucleotide cleaved off from n at its 3'-end.

For identification of consecutive degradation products after electrophoresis it is necessary to sort these in a second chromatographic step according to their chain length. The material on the electrophoresed cellulose acetate strip is transferred to the start line of a DEAE-cellulose thin layer plate. The plate is developed with a partial hydrolysate of RNA in the mobile phase. The basis of the separation is a co- or homochromatography of the radioactive degradation products with non-labelled RNA-oligomers of the same length. The separation of the RNA mixture proceeds via displacement of short oligomers by longer oligomers having more negative charges and which bind stronger to the DEAE-cellulose: That means short oligomers migrate faster than the longer ones. A further sequence information is obtained from R_f differences in homochromatography. The cleavage of a purine nucleotide can be clearly distinguished from that of a pyrimidine nucleotide, since ΔR_f is greater for purine cleavages.

The combined effects from electrophoresis and homochromatography on the mobility shifts in a 2D-fingerprint are shown in figure 3.

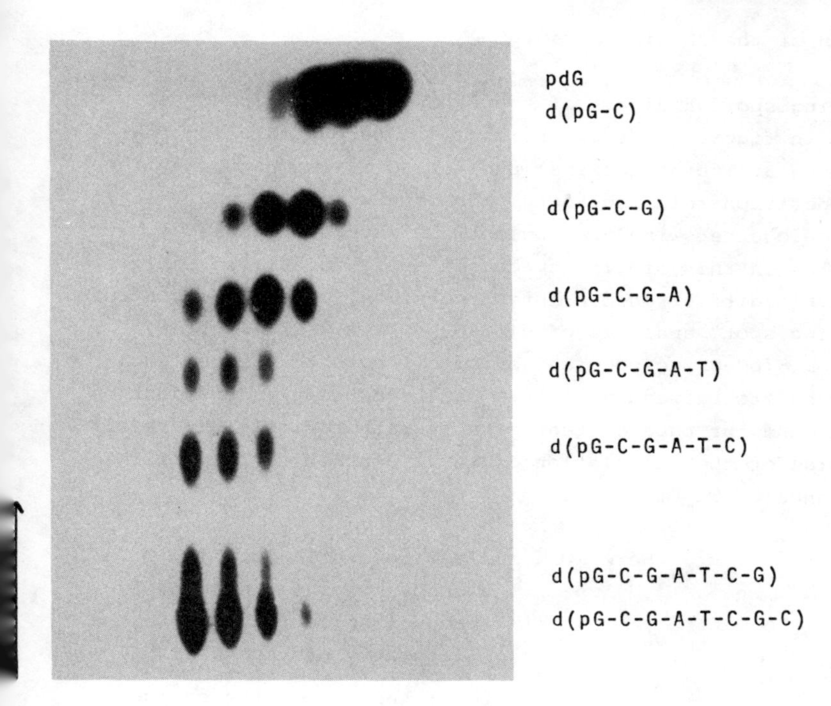

Figure 1 Homochromatography of the partial SVPD-hydrolysis of d(pG-C-G-A-T-C-G-C)

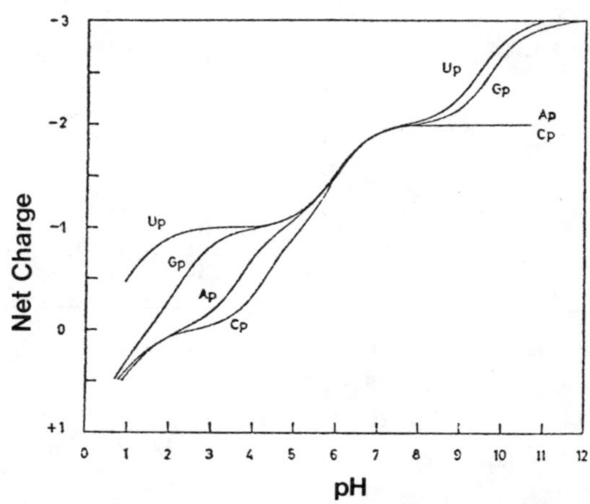

Figure 2 Net charge of the 3'-ribonucleotides as a function of the pH value [7]

1.2 Evaluation of the 2D-fingerprints

The "wandering spot" analysis of the octanucleotide d(pG-C-G-A-T-C--G-C) is shown in figure 4. It can be easily interpreted using the rules from figure 3. The 5'-terminal nucleotide is unambiguously identified by its position relative to the blue dye xylene cyanol that is chromatographed together with the nucleotidic material (table 1).

However, rules in this simplicity do not apply for all possible sequences and may lead to erroneous interpretations. This is demonstrated in the "wandering spot" analysis of d(pC-C-T-G-C-A-G-G) (Fig. 5). The C-shift from d(pC-C) to pdC is just opposite to that expected. Thus, mobility shifts between n and n-1 can be extremely dependent on both the length and the base content of n. A rather simple but fairly accurate formula for the calculation of the electrophoretic mobilities has been published by Tu et al. [3].

$$U_{XC} = \frac{Q_n}{K_n}$$

where U_{XC} = electrophoretic mobility of a 5'-phosphorylated oligodeoxyribonucleotide relative to the blue dye xylene cyanol (XC)

Q_n = Σq_{XC}; the sum of the charge equivalents of the component mononucleotides

K_n = length dependent retardation coefficient

Values for q_{XC} are given in table 1 and those for K_n in table 2.

Table 1

mononucleotides	pKa	q(calc.)	q(obs.) [a]	q_{XC}(obs.) [a]
pdT	-	-1.00	-1.00	-1.27
pdG	2.9	-0.80	-0.79	-1.00
pdA	3.8	-0.34	-0.26	-0.33
pdC	4.6	-0.07	-0.05	-0.06

[a] The q values are determined from the average mobility for each of the mononucleotides

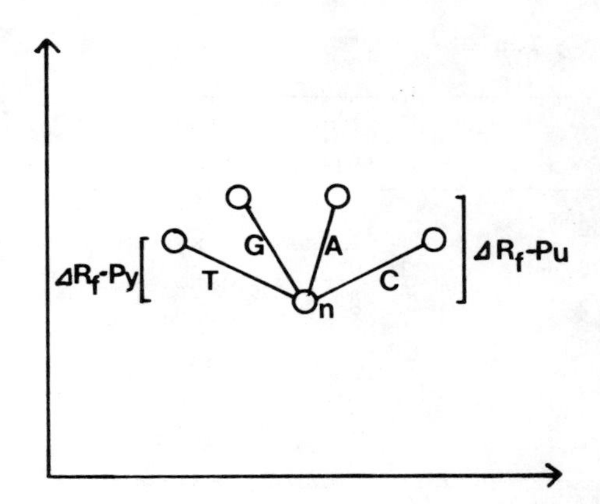

Figure 3 Characteristic mobility shifts between n and n-1

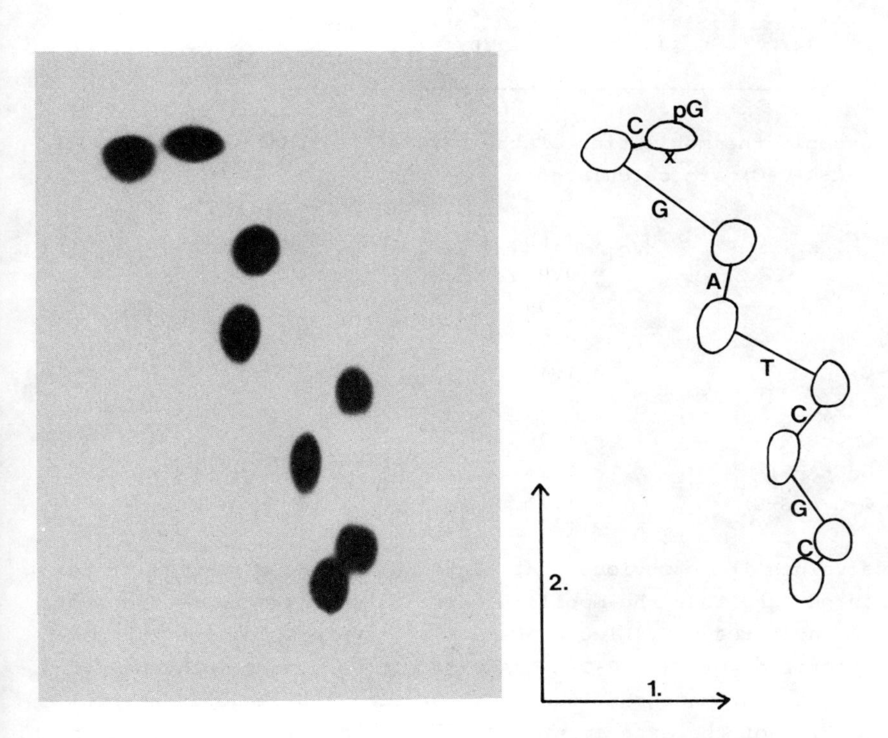

Figure 4 Fingerprint of d(pG-C-G-A-T-C-G-C)
1. hve; 2. homochromatography; x = position of XC

Table 2

n	$\overline{K}_n$	relative deviation [%]
1	1.00	7.8
2	1.30	10.0
3	1.72	5.8
4	2.02	6.4
5	2.40	6.3
6	2.76	5.1
7	3.06	5.9
8	3.37	4.7
9	3.65	3.6
10	3.99	4.8
11	4.31	3.9
12	4.68	3.6
13	5.09	2.4
14	5.34	3.0

As an example the mobilities of the partial degradation products of d(pC-C-T-G-C-A-G-G) are calculated:

XC	U_{XC} = 1.00
pdC	= 0.06
d(pC-C)	= 0.09
d(pC-C-T)	= 0.79
d(pC-C-T-G)	= 1.17
d(pC-C-T-G-C)	= 1.01
d(pC-C-T-G-C-A)	= 1.00
d(pC-C-T-G-C-A-G)	= 1.23
d(pC-C-T-G-C-A-G-G)	= 1.41

From these values it is obvious that the dinucleotide d(pC-C) migrates faster than pdC and that the mobility shift between the upper two spots in figure 5 indicates a C. Figure 6 shows a computer plot for the calculated fingerprint of d(pC-C-T-G-C-A-G-G). It is almost identical to the experimental chromatogram. Such a theoretical 2D-image is helpful in the selection of the experimental conditions and in the interpretation of the results.

The "wandering spot" technique is still the method of choice for the sequence analysis of shorter oligonucleotides ($n \leq 40$). There

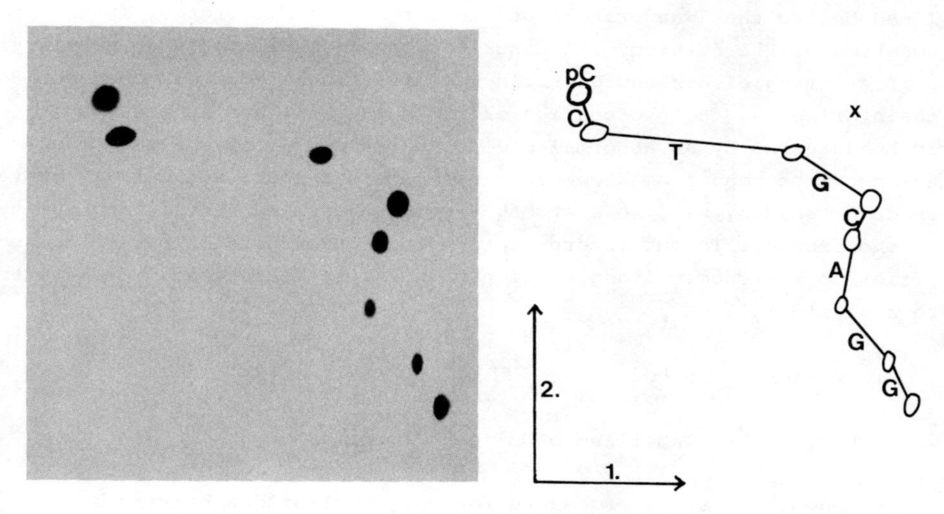

Figure 5 Fingerprint analysis of d(pC-C-T-G-C-A-G-G)

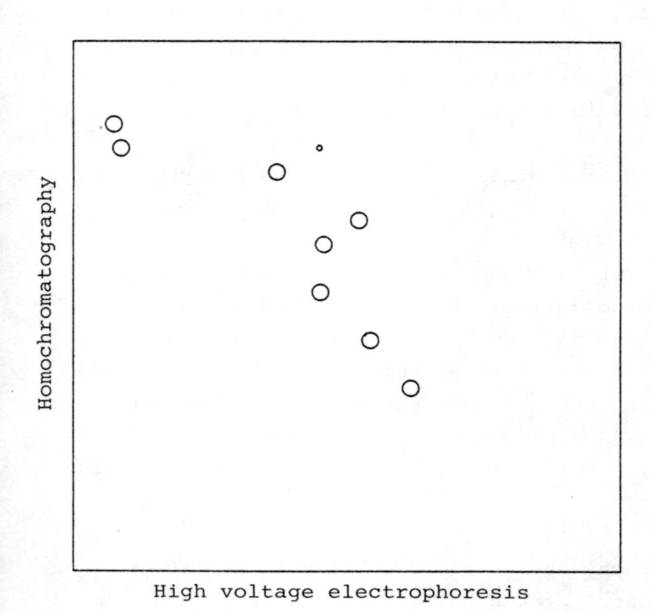

Figure 6 Computer plot of the calculated fingerprint of d(pC-C-T-G-C-A-G-G) [3]

is an unique feature that makes this method very attractive. As outlined before the "wandering spot" relies on the differences in the protonation of the heterocyclic bases leading to characteristic mobility shifts during electrophoresis. Any modification of the oligonucleotide chain affecting the protonation of the bases will easily and sensitively be detected by an abnormal electrophoretic mobility. A sequence which can be read from a complete set of clear spots without any extra spots or shades, will be a strong argument for the identity and purity of the product. To our interpretation the wandering spot method is superior to the chemical degradation when short oligonucleotides have to be analysed [4].

2 Discussion of methods

2.1 The [5'-^{32}P]-labelling of oligonucleotides

The phosphorylation procedure for oligonucleotides bearing a free 5'-OH function using [γ-^{32}P]ATP and T4 polynucleotide kinase is described in this volume (see Fritz, Gait, Uhlenbeck).

After labelling the oligonucleotide is separated from excess of [γ-^{32}P]ATP and contaminating ^{32}P-phosphate. Procedures such as Sephadex G10 filtration, chromatography or DEAE-paper, hplc, polyacrylamide-gel electrophoresis (page), may be used. We support the use of the page procedure described in detail by Gait (this volume).

2.2 Hydrolysis of the oligonucleotides by snake venom phosphodiesterase (SVPD)

The reaction conditions are designed to result in a slow degradation of the labelled oligonucleotide. This is achieved by addition of a large excess of unlabelled RNA. The enzyme activity, as applied in a defined ratio over carrier RNA will completely degrade the RNA within appr. 4 h. Thus independent of its length the neglectable amount of labelled oligomer is digested during the same period of time.

2.3 Homochromatography

The term homochromatography designates a thinlayer chromatography (tlc) on DEAE-cellulose coated plates performed at 60°C with an RNA hydrolysate (homo-mix) in the mobile phase. The mobile phase also contains 7M urea to suppress any hydrogen bond interaction. Due to this high content of urea care must be taken that the front of the mobile phase will not become dry. This would cause crystallisation of urea and

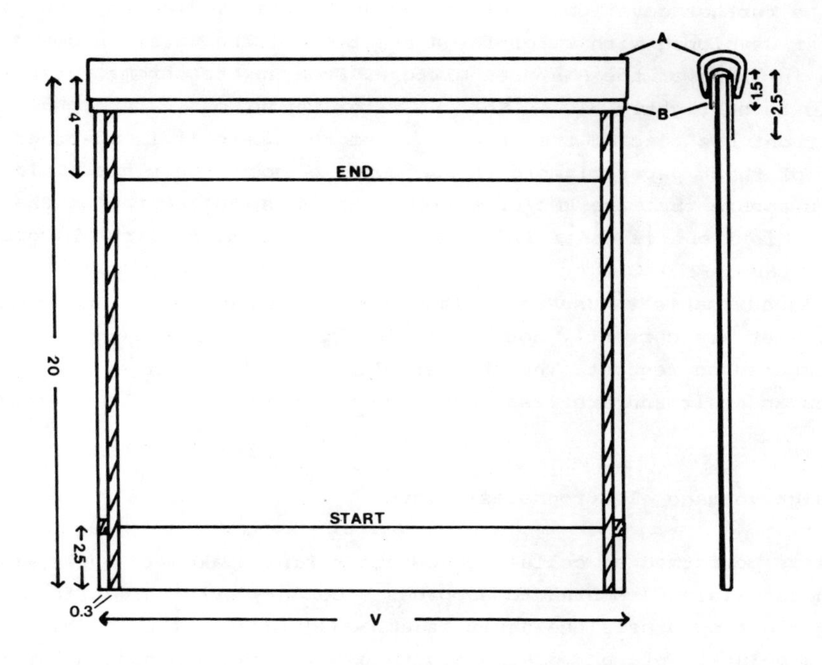

Figure 7 Design of a DEAE-cellulose plate for homochromatography. All distances are given in cm; //// coat removed; A: plastic clip; B: 3 MM paper; V: variable

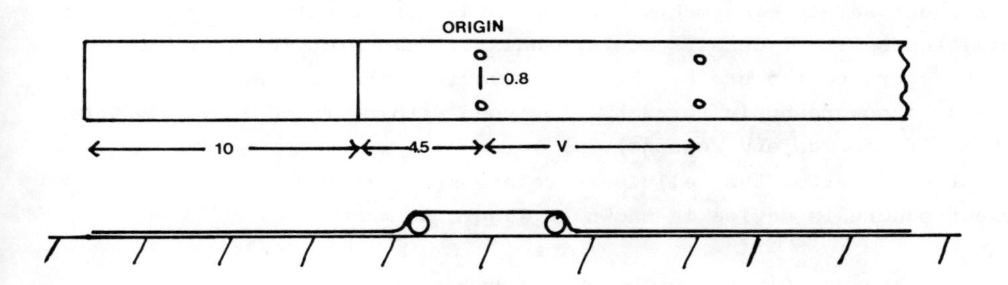

Figure 8 Design of the cellulose acetate strip. All distances in cm; V: variable o: start and end points of XC

stop the further development of the chromatogram. Therefore, the plate is first developed with water for a short time. The water chromatographs in front of the RNA/urea mixture. Even so the chromatographic tank is to be tightly closed that no water vapour can escape. When the water front has reached the upper edge of the plate it is adsorbed by a wick of thick paper clambed to the upper edge of the plate. This procedures ensure that the RNA/urea front can reach to the top of the plate. This front is controlled visually by migration of the blue dye xylene cyanol FF (XC).

It is advisable to use only glass plates with coats at least 0.2 mm thick. They are currently not available from commercial suppliers, but are prepared on request. The thicker plates can hold larger amounts of samples and salt and are less sensitive to drying during chromatography.

2.4 High voltage electrophoresis (hve)

hve is performed on cellulose acetate strips soaked in buffer of pH 3.5 containing 7 M urea to suppress secondary structure effects. During electrophoresis the strip is embedded in petroleum of high boiling point to prevent drying and to provide uniform cooling. Besides the commercially available huge hve devices (Savant, Gilson) containing about 100 L of petroleum, two smaller models are described in literature [3,5]. The one used by us [5] is fairly small (constructed into a chromatographic tank for 20 cm x 20 cm plates), handy and safe. It contains only 3 L of petroleum and can be installed without very stringent safety equipment. The apparatus can be obtained from commercial sources; for details see appendix 1. This protocol is written with regard to the use of the latter device. The apparatus is connected to a power supply for 3 kV, equipped with a ground fault leakage shut off (for safety reasons).

Attachement of the cellulose acetate strip to the rack of the electrophoresis device is shown in figure 9.

2.5 On how to select the 2D-conditions

The 2D-procedure can be devided into three steps: high voltage electrophoresis, transfer, and homochromatography. Together with the partial digest of the radioactively labelled oligonucleotide the blue dye xylene cyanol FF is also submitted to these procedures. The dye serves as a visual marker during all operations of the fingerprinting. It helps to adapt the experimental conditions to the sequence you have

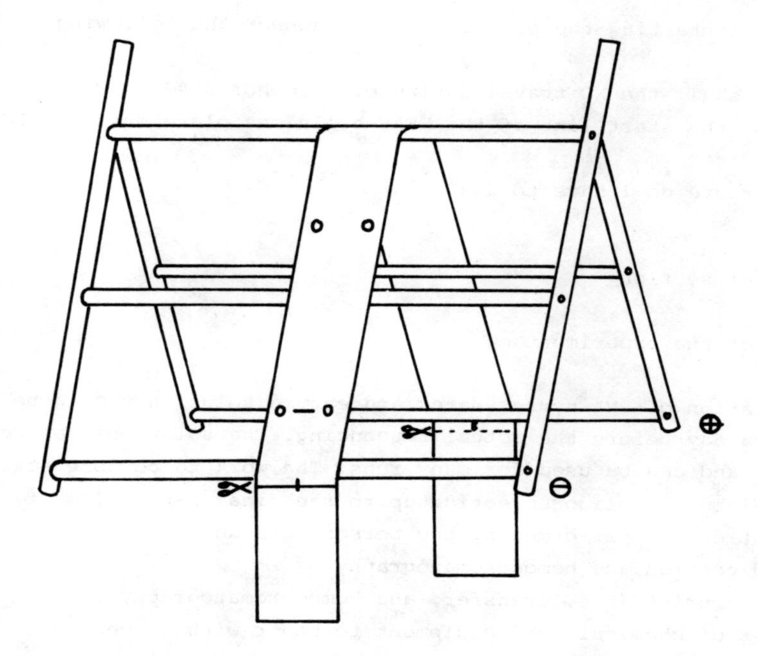

Figure 9 Fixing of the cellulose acetate strip to the rack of the hve apparatus

to analyse. Two parameters can now be varied to spread the "wandering spots" over the 20 cm x 20 cm surface of the homochromatogram: 1. duration of electrophoresis; 2. degree of hydrolysis of the homo-mix RNA.

The first parameter has to be varied considerably according to the actual base sequence. Thus, it is advisable to calculate the expected relative mobilities for all components of the partial digest using the formula and values described in 1.2. For a given distance (e.g. 10 cm) the XC shall migrate during electrophoresis, calculate the mobilities (cm) for the slowest and the fastest component. Notice that XC is regarded as additional component of the partial digest. Since only 18.8 cm of the tlc start line is available for the transfer, the difference between the slowest and the fastest moving components should not exceed 15 cm. A more convenient way to determine the duration of electrophoresis is to use the computer plots for various XC-distances.

The second parameter, hydrolysis of RNA for the mobile phase, has to be varied only in coarse steps according to the length of your oligonucleotide, and the choice of the homo-mixture to be used is rather uncritical.

Thus, before starting the practical work, answer the following questions:
What distance shall the XC travel during electrophoresis?
Which point on the start line of the DEAE-cellulose plate must the XC be transferred to?
Which homo-mixture do I have to use?

3 Experimental section

3.1 Outline of the experiment

The preparation of RNA hydrolysates and enzyme solution should be started half a day before the actual sequencing. The solutions can be kept on stock and can be used for many runs. The work to be done starting with the labelled oligonucleotide up to the final X-ray film of the 2D-fingerprint can be performed by one person in 2 days.
Day 1: SVPD digestion and homochromatography
Day 2: 2D-fingerprint (hve, transfer, and homochromatography).

A full list of chemicals and equipment together with suggested suppliers is given in appendix 1.

3.2 The labelled oligonucleotide

The purified, lyophilized [5'-^{32}P]-oligonucleotide (10^5 cpm (Čerenkov) are sufficient) is dissolved in 40 µL of 10 mM Tris-HCl, pH 8.0 and stored at -20°C.

3.3 General performance of the "homochromatography"

Two glass tanks for chromatography (to fit with 20 cm x 20 cm plates) with well greased ground cover are thermostated to 60°C in a drying oven preferably equipped with a glass door. 60 mL of water is added to one tank.

A DEAE-cellulose plate is prepared as shown in figure 7. The start and end lines are marked with a pencil. Between two sets of lines, 3 and 6 mm parallel to the left and right edge of the chromatogram, the cellulose coat is removed with a spatula. A 4 cm x V cm strip of 3 MM paper is folded lengthwise (1.5 cm - 2.5 cm) and clambed to the top of the plate using a plastic clip.

After application of the samples, 3 to 4 small aliquots of XC-solution are added along the start line. This serves as a visual marker to control the development of the chromatogram.

60 mL of a homo-mixture are added to the second tank. The tlc plate is placed into the water tank. When the water front has travelled 5 to 7 cm the plate is quickly transferred to the second tank. During the first minutes take care that the increasing vapour pressure does not lift the cover and that the tank remains tightly closed. When the blue dye has reached the marked line (∿2.5 h) take out the plate, remove the 3MM paper and dry the plate immediately under a stream of warm air.

Apply some markers of radioactive ink asymmetrically to the bottom part of the plate (below starting line), cover the plate with plastic wrap and expose to an X-ray film in combination with an intensifying screen.

3.4 Snake venom phosphodiesterase hydrolysis

Prepare a DEAE-cellulose plate (8 cm x 20 cm) as described under 3.3 and mark with a pencil 6 points in 1 cm distance along the starting line. Mark six 0.4 mL plastic tubes with numbers 1 to 6 (use a water-resistant pen).

Draw out a glass capillary (original Ø = 1.5 mm) and calibrate with H_2O to hold 6 µL.

Prepare a Dewar vessel with liquid nitrogen or lumbs of dry ice.

Set one water bath to 37°C and another one to 95°C.

The enzymatic hydrolysis is performed in a 1.5 mL Eppendorf tube. Combine:

20 µL	oligonucleotide solution (see 3.2)	
4 µL	carrier-RNA (see appendix 2)	
5 µL	SQ-mix	" " "
20 µL	water	
10^{-3} U	SVPD	" " "
∿50 µL total volume		

Start the reaction by adding the enzyme and incubate at 37°C. After 10, 20, 40, 90, 180, and 360 min remove aliquots of 6 µL from the reaction mixture with the calibrated capillary. Apply appr. 1 µL to one point on the start line of the DEAE-cellulose plate and blow out the remainders to the bottom of a microtube. Close the microtube and heat to 95°C for 1 min to inactivate the enzyme. Then throw the microtube into the Dewar vessel.

After removal of the last aliquot the Eppendorf tube is kept at room temperature overnight and the final portion of the reaction mixture is inactivated at 95°C the next day. Saving this sample may be advantageous if the enzyme was not sufficiently active. The six microtubes are stored at -20°C.

The DEAE-cellulose plate is subjected to homochromatography and exposed to an X-ray film overnight.

3.5 The 2D-fingerprint

The high voltage electrophoresis and the transfer to the tlc plate have to be performed without interruption.

3.5.1 The first dimension

Set up electrophoresis device and fill with fresh hve-buffer (50 mL) each reservoir.

Clean a 1 m x 1 m plane working space (preferably a plastic coated non-flexible plate) and fix two glass pipettes (10 mL) in parallel at a distance of 5 cm with adhesive tape. They will support the cellulose acetate strip during the application of the "cocktail" (see figure 8). Fill a small bowl with hve-buffer.

Wrap two short test tubes (Ø ∿1.5 cm) with tissue paper. Provide a pack of tissue paper, a roll of plastic wrap and scissors.

Draw out two glass capillaries (one for the cocktail, and one for XC-solution).

Take a cellulose acetate strip and mark with a ballpoint pen as shown in figure 8 (Do not touch the strip with bare fingers!). Preparation of the "cocktail": the X-ray film of the homochromatogram as obtained in 3.4 is used to calculate the composition of a "cocktail" (3-4 µL) containing all partial degradation products to an equal amount. Derived from figure 1, e.g. the composition of such a "cocktail" would be as follows: 2 µL of aliquot 2 and 2 µL of aliquot 4. The calculated volumes are withdrawn from the microtubes using a calibrated (1-5 µL) glass capillary and transferred to the home-made capillary. The "cocktail" is now ready for application to the cellulose acetate strip.

The cellulose acetate strip is soaked in hve-buffer by slowly passing it through the bowl. Place the strip across the two glass pipettes with the marked start line in-between. Further wet the strip intensively with tissue paper soaked in hve-buffer. Take the two wrapped short tubes and place one below and the other above the start line. Press together for some sec so that this region of the cellulose acetate becomes fairly dry. Remove the wrapped tubes, re-adjust the strip and cover the protruding ends left and right of the pipettes with plastic wrap. Next the "cocktail" is slowly applied to the start line by gently blowing out the solution from the capillary and waiting each time for the drop on the surface to be sucked up by the cellulose acetate. After application of the "cocktail" place two small drops of XC-solution into the marked circles left and right from the sample on the start line. Remove the plastic wrap and gently plot excess buffer liquid with tissue paper. Fix the strip onto the angled rack of the electrophoresis device as outlined in figure 9 (the start line is placed 4.5 cm away from that end of the rack that will dip into the cathode (-) buffer reservoir). Cut off excess parts of the strip and insert the rack slowly into the electrophoresis tank. All operation from the application of the "cocktail" up to insertion into the tank should not take longer than 10 min in order to avoid drying of the cellulose acetate strip. Electrophoresis is routinely performed at 2.5 kV (80 V/cm) until the XC reaches the marked circles; the XC migrates with a velocity of ∿8 cm/h.

3.5.2 The transfer from the cellulose acetate strip to the tlc plate (Fig. 10)

A 1 mL plastic pipette (use only the type described in appendix 1) and a 1 mL glass pipette are fixed in parallel and at a distance of 10 cm with adhesive tape to the working bench.

Cut out 4 strips of 3 MM paper (4 cm x 25 cm) and wet them intensively with water. Prepare a DEAE-cellulose plate (20 cm x 20 cm) as described in 3.3. In addition the left and right end points of the start line are marked by removal of the cellulose coat so that they can be identified by looking from the back of the plate. The position where the XC is to be transferred to is marked with a water-resistent pen on the back of the plate.

When the electrophoresis is finished, remove the rack from the tank and place it on paper towels to suck up the dripping petroleum. Take off the cellulose acetate strip and place it on the plastic pipette with the origin of electrophoresis to your left hand and the portion to be transferred in the center. Place two of the wet 3 MM strips along each endge of the cellulose acetate strip with only a 1-2 mm overlap. Carefully put the DEAE-cellulose plate, with the coat downwards, onto the glass pipette, and position the plate such that the start line is adjusted accurately upon the pipette and the marked position for XC coincides with the blue dye on the cellulose acetate strip. Layer the plate horizontally onto the plastic pipette and load it with a weight of ∿1 kg. The liquid from the cellulose acetate strip and the paper

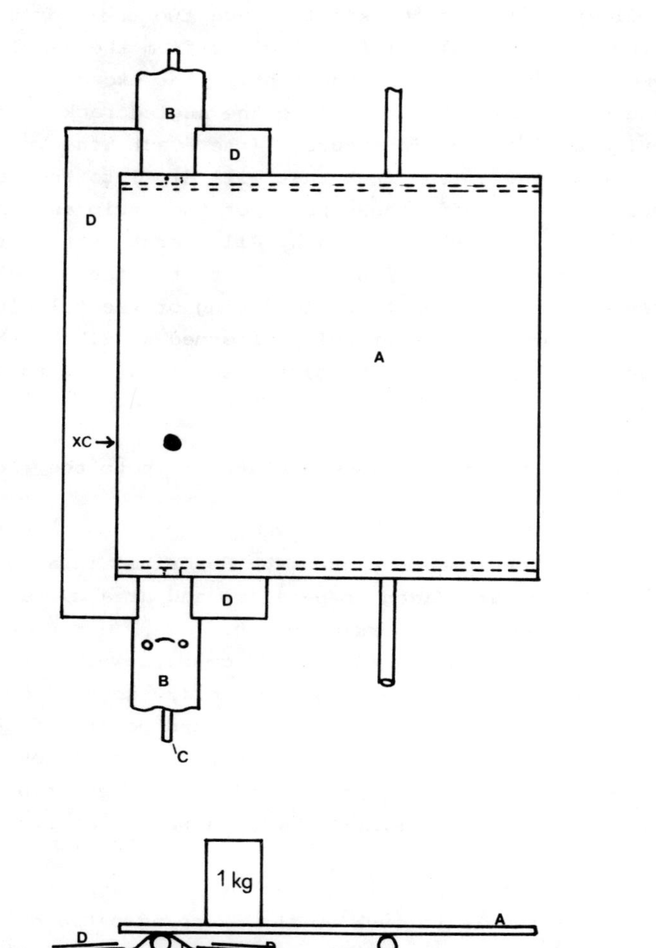

Figure 10 Details of the transfer experiment
A: DEAE-cellulose plate; B: cellulose acetate strip;
C: serological pipette; D: 3 MM paper strips;
•: marker for XC; ----- seen through the glass plate

wick is now sucked up by the DEAE-cellulose coat. More than 80% of the radioactive material is transferred within 30 min. Remove the plate and dry it under a stream of cold air. Rinse the plate with 96% ethanol (technical grade), to remove the transferred urea and dry again.

3.5.3 Second dimension

The DEAE-cellulose plate is subjected to homochromatography as described in 3.3.

REFERENCES

[1] Brownlee, G.G. and Sanger, F. (1969) Eur. J. Biochem. 11, 395

[2] Jay, E., Bambara, R., Padmanabhan, R., and Wu, R. (1974) Nucl. Acids Res. 1, 331

[3] Tu, C.-P.D., Jay, E., Bahl, C.P., and Wu, R. (1976) Anal. Biochem. 74, 73

[4] Maxam, A.M. and Gilbert, W. (1977) Proc. Natl. Acad. Sci. USA 74, 560

[5] Blöcker, H. and Köster, H. (1978) Liebigs Ann. Chem. 1978, 982

[6] Silberklang, M. and Gross, H.J., pers. communication

[7] Organic Chemistry of Nucleic Acids, Part A; Kochetkov, N.K. and Budovskii, E.I., eds.; Plenum Press, London, New York

Appendix 1: List of equipment

A Homochromatography

	Item	Suggested Supplier
1.	tlc-glass plates (20 cm x 40 cm) coated with CEL DEAE/HR-Mix 20 (0.2 mm), pack of 25	Macherey and Nagel (Specially prepared when ordering 3 packs)
2.	2 tanks for tlc, No. 120 160	Desaga
3.	drying oven with glass door	Heraeus, Memmert
4.	paper for chromatography 3 MM CHR, No. 3030917	Whatman
5.	plastic wrap (Saranwrap)	
6.	1 glass knife	
7.	1 spatula (3 mm wide)	
8.	plastic clips (20 cm length)	office supplies
9.	1 hair dryer	
10.	X-ray sensitive films + intensifying screen	
	Ortho G + Lanex Regular or	Kodak
	CURIX RPI/100 AF4 + CURIX MR600	Agfa
11.	X-ray film developer and fixative	Kodak
12.	trays for film development	
13.	X-ray film developing facilities	
14.	wash-bottle filled with ethanol (technical grade)	

B SVPD Hydrolysis

	Item	Supplier
1.	1.5 mL Eppendorf tubes	
2.	0.4 mL polyethylene microtubes (uncoated) No. 314 326 pack of 1000	Beckman
3.	water-resistant pen	
4.	glass capillaries, Ø = 1.5 mm, length 100 mm	
5.	micropipettes (1-5 µL), No. 708 707	Brand
6.	1 set of adjustable microliter pipettes with appropriate tips	Eppendorf or Gilson
7.	1 water bath at 37°C	
8.	1 water bath at 95°C	
9.	1 stop watch	

10. 1 Dewar vessel
11. liquid nitrogen or dry ice

C High voltage electrophoresis (hve)

No.	Item	Supplier
1.	1 hve apparatus: "Miniatur-Hochspannungs-Elektrophorese-Apparatur" (model without cooling coil)	H. Hölzel-Technik, Bernöderweg 7, D-8250 Dorfen 1
2.	5 L "Varsol" 155/185 petroleum	Esso (ask for a free sample)
3.	cellulose acetate strip for electrophoresis CA 250/0 (30 mm x 550 mm) No. 480014, pack of 50	Schleicher & Schüll
4.	1 high voltage power supply with ground fault leakage shut off	Pharmacia
5.	1 m x 1 m plastic-coated inflexible plate (lab bench)	
6.	glass pipettes (10 mL)	
7.	adhesive tape (Tesa film or Scotch Magic)	
8.	1 glass bowl (50 mL volume)	
9.	tissue paper, No. 6017	Kleenex
10.	2 short test tubes Ø ~1.5 cm	
11.	scissors	

items A5, B4, and B5 are also required

D Transfer

No.	Item	Supplier
1.	1 serological pipette (1 mL) No. 7521F	Falcon
2.	1 glass pipette (1 mL)	
3.	1 weight of appr. 1 kg	

items A1, A4, A9, B3, C5, and C7 are also required

E Chemicals and equipment used in appendix 2

Standard chemicals such as buffer salts, urea etc. are of highest purity available e.g. p.a. Only double distilled water is used.

No.	Item	Supplier
1.	RNA (yeast) No. 109 223	Boehringer
2.	bis-(4-nitrophenyl)-phosphate (sodium salt) No. 15118	Serva
3.	xylene cyanol FF, No. 38505	Serva
4.	4 polyethylene bottles (1 L)	

5. 4 polyethylene bottles (0.1 L)
6. 10 polyethylene tubes (10 mL)
7. 1 beaker (1 L)
8. 1 beaker (0.1 L)
9. Teflon stirring bars (small and large)
10. 1 measuring flask (1 L)
11. 1 measuring flask (0.1 L)
12. 1 measuring flask (10 mL)
13. 1 magnetic stirrer with heating plate
14. big water bath set at 65°C (cooking pot)
15. a few Pasteur pipettes, glass
16. pH-Meter
17. snake venom phosphodiesterase (EC 3.1.4.1) No. 3926 — Worthington
18. uv/vis-spectrophotometer with thermostatable cell-holder — Zeiss, Beckman etc.
19. Parafilm (M) — American Can Comp.

Appendix 2:

F Preparation of the solutions

1. 2 M KOH
   Dissolve 11.2 g KOH in 70 mL water and fill up to 100 mL. Store tightly closed in a polyethylene bottle.
2. 10 M KOH
   Dissolve 56.1 g KOH in 70 mL water and fill up to 100 mL. Store tightly closed in a polyethylene bottle.
3. RNA-hydrolysate for homochromatography (homo-mix) [6]
   Specifications: 3% (w/v) yeast RNA hydrolyzed with X mM KOH, 7 M urea, pH 6.0.
   Weigh out 420 g urea and 30 g RNA in a 1 L beaker.
   Dissolve in 400 mL water by stirring and keeping at 20°C.
   Adjust the pH to 7.0 with 2 M KOH and fill up to 0.9 L with water. Pour into a 1 L measuring flask. Arrange a water bath (big enough to take the measuring flask) on a magnetic stirrer and heat to 65°C.
   Add the appropriate volume of 10 M KOH to the flask:
   7.5 mL for sequencing 4-8mers, final conc. X = 75 mM

5.0 mL for sequencing 8-12mers, final conc. X = 50 mM
3.0 mL " " 12-16mers, " " X = 30 mM
Place the flask into the water bath and incubate with stirring for 20 h. After cooling to room temperature (tap water) pour back into the beaker and adjust the pH to 6.0 with glacial acetic acid. Fill up to 1 L and store in a polyethylene bottle at 4°C.

4. Carrier RNA
Specification: 20 mg/mL hydrolyzed yeast RNA, pH 7.0.
Dissolve 2 g RNA in 5 mL water and transfer the solution into a 10 mL measuring flask. Add 1 mL 10 M KOH, fill up to 10 mL with water and incubate at 37°C for 24 h. Then dilute to appr. 80 mL with water, neutralise with 1 N HCl and fill up to 100 mL. Store in a polyethylene bottle at -20°C.

5. SVPD - test solution
Dissolve 0.36 g bis-(4-nitrophenyl)-phosphate in 20 mL 1 M Tris--HCl, pH 9.0. Dilute to 100 mL with water and store in a polyethylene bottle at -20°C.

6. SVPD - solution
Dissolve the lyophilized enzyme powder in water to a concentration of 1 $mg \cdot mL^{-1}$. Devide the solution into 0.1 mL aliquots and store in 0.4 mL microtubes at -20°C. The aliquot in use is kept at 4°C.
Activity test: 1 U is the amount of enzyme that liberates 1 µmol 4-nitrophenol from bis-(4-nitrophenyl)-phosphate at pH 9.0 and 37°C. Fill two uv-cells, each with 1 mL of SVPD-test solution (E 5) and place them into a spectrophotometer with the cuvette holder thermostated to 37°C. Add an enzyme aliquot (V = 1 to 5 µL) to the sample cell, mix and read the absorbance at 405 nm every min. The activity is calculated from an average $\Delta A_{405} \cdot min^{-1}$ using the following formula:

$$\frac{U}{mL} = \frac{\Delta A_{405} \cdot 10^3}{18.5 \cdot V\ [\mu L]}$$

7. SQ-mix (10X)
Prepare 10 mL containing 0.5 M Tris, 0.05 M potassium phosphate, pH 8.6, and store in a polyethylene bottle at -20°C.

8. hve - buffer
Specification: 7 M urea, 5 mM EDTA, pH 3.5.
Dissolve 420 g urea and 0.93 g $EDTA \cdot Na_2$ in 700 mL water and adjust the pH to 3.5 with glacial acetic acid. Fill up to 1 L and store

in a polyethylene bottle at 4°C.

9. xylene cyanol solution
   Specifications: 1% dye in ethanol:water (1:1, v/v)
10. radioactive ink
   Mix the xylene cyanol solution with some radioactive material e.c. $[^{32}P]H_3PO_4$ or $Na_2[^{35}S]SO_4$.

Index